Thomas Workman

Malaysian Spiders

Thomas Workman

Malaysian Spiders

ISBN/EAN: 9783337295080

Printed in Europe, USA, Canada, Australia, Japan

Cover: Foto ©berggeist007 / pixelio.de

More available books at **www.hansebooks.com**

SPIDERS.

BY

THOMAS WORKMAN,

Vice-President Belfast Natural History and Philosophical Society.

VOL. I.

MALAYSIAN SPIDERS.

PUBLISHED BY THE AUTHOR.

BELFAST.

1896.

PREFACE.

In compiling the volume on Spiders now laid before the public, the Author's endeavour has been to aid future workers in this interesting subject by supplementing with drawings the well-known works of Professor T. Thorell, of Montpellier, to whom he owes a great debt of gratitude for continuous favours during many years. The first twenty-four drawings, in which the Author was assisted by his daughter, who unfortunately had to discontinue her help owing to her eyesight becoming impaired, were brought out in three parts—part 1 in 1892, and parts 3 and 4 in 1894—under the title of Malaysian Spiders.

Nearly all the specimens were collected by the Author in Singapore and Java, and are in his possession except where otherwise noted.

The male is denoted by ♂, and the female by ♀, the distinctive marks usually adopted in works of this nature. The drawings of the palpi and epigynae were made when the parts were dry except when stated otherwise, but the colours have been taken from specimens as they appear after immersion in spirits of wine, and not from living spiders.

The Author hopes the notes on the habits and webs of the various species will be of interest, so little being known of the life history of tropical spiders.

> "The heart is hard in Nature,
> ———that is not pleased
> With sight of animals enjoying life,
> Nor feel their happiness augment his own."
>
> —Cowper.

Craigdarragh,
Co. Down,
31st August, 1896.

ALPHABETICAL INDEX.

ALPHABETICAL INDEX.

ADDENDA.

Page 3—*P. rostratus*, Thor., is a synonym of *Strigoplus albo-striatus*, Sim. Bull. Soc. Zool., France, p. 446. 1885.

„ 4—*C. formosus*, Thor., is a synonym of *Camaricus striatipes*, Van Hass. Midden Sumatra, etc., Aran., p. 43, pl. iii., figg. 7 and 8. 1882.

„ 16—*H. virens*, Thor., is a synonym of *Ergane sannio*, Thor. The Spiders of Burma, p. 391. 1895.

„ 17—*A. sumatranus*, Thor., is a synonym of *Argyrodes flavescens*, Cambr. Proc. Zool. Soc., London, p. 321, pl. xxviii., figg. 1-1f. 1880.

Plate 97—*Ocyale hirsuta*, Work., should be out.

OXYOPES LINEATIPES. C. L. Koch.

Syn.: 1848. *Sphasus lineatipes*, C. L. Koch, Die Arach. xv., p. 55, tab. DXVIII., fig. 1455.

1891. *Oxyopes* „ Thor., Spindlar fran Nikobarerna och andra delar af Södra Asien. (K. Svenska Vetenskaps—Akademiens Handlingar, XXIV., No. 2), page 71.

Description of Plate 1 ♂—*a*, figure of spider magnified; *b*, natural size; *c*, front view of eyes and falces, (eyes of first and second row too far apart in drawing); *d*, epigyne; *e*, profile of spider; *f*, right palpus of ♂.

Total length, ♂ 7 m.m. ♀ 9 m.m.; cephalo thorax, ♂ 3 m.m. ♀ 3 m.m.; abdomen, ♂ 4 m.m. ♀ 6 m.m.

Length of legs, ♂ i.—15, ii.—14, iii.—11, iv.—15·5 milemetres.
„ ♀ i.—13, ii.—12, iii.—10, iv.—13 milemetres.

The eyes in the front row are the smallest. Those of second row about the diameter of one of them apart. Clypeus rather higher than width of second row of eyes.

Four black lines on upper side of abdomen, and black line on lower side of first pair of legs.

Epigyne horse shoe shaped, convex side backwards. Spider when alive of a brilliant green colour, and is to be found in abundance in Singapore on bamboo hedges. It does not seem to make any web, but runs freely over the leaves. Very common in the Malay Archipelago.

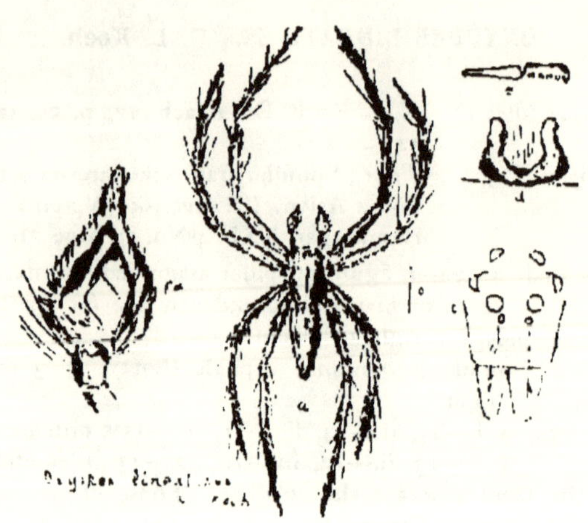

OXYOPES GEMELLUS. Thor.

Syn. : 1891. *Oxyopes gemellus*, Thor., Spindlar fran Nikobarerna och andra delar af Södra Asien. (K. Svenska Vetenskaps—Akademiens Handlingar, XXIV., No. 2), page 71.

Description of Plate 2 ♂—*a*, figure of spider magnified; *b*, natural size; *c*, front view of eyes and falces; *d*, profile of spider; *e*. left palpus.

This spider, found in Penang by myself in 1888, is in all respects exactly similar to *Oxyopes lineatipes*, C. L. Koch, except that the digital joint of the palpus is more prolonged at the joint, and a slight difference in the shape of the palpal organs. It may, therefore, not improbably turn out that this is but a variety of *O. lineatipes*, C. L. Koch. The ♀ has not yet been found.

PELTORHYNCHUS ROSTRATUS. Thor.

Syn.: 1890. *Peltorhynchus peltatus*, Thor., Aracnidi di Pinang p. 11.
1891. do. *rostratus*, id., Spindlar fran Nikobarerna och andra delar af Södra Asien. (K. Svenska Vetenskaps—Akademiens Handlingar, XXIV., No. 2), p. 88.
1891-92. do. do. id., Studi, cet., IV., Ragni dell' Indo-Malesia, cet., II., p. 114.

Description of Plate 3 ♀—*a*, figure of spider magnified; *b*, natural size; *c*, epigyne; *d*, labium and maxillae; *e*, eyes and falces; *f*, profile of spider.

Labium and maxillae covered with short spines. Strong spines on the front of the falces. Row of short strong spines on the edge of the clypeus. Long strong spines on the ocular area, and over the cephalothorax.

The genus *Peltorhynchus*, Thor., can easily be distinguished by the peculiar projecting clypeus. ♂ found in Java by Professor Kinberg, and ♀ by me in Penang.

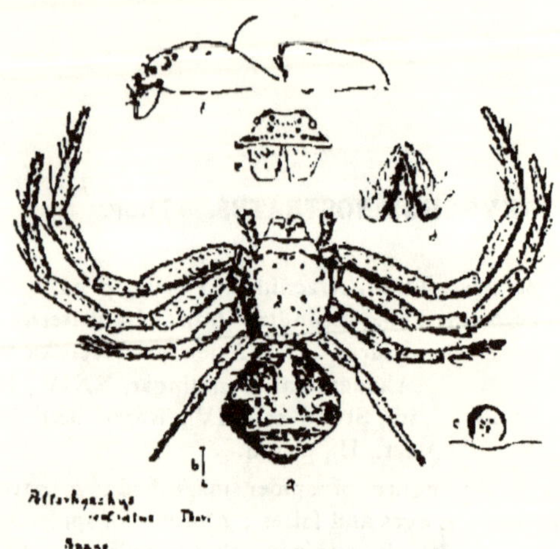

CAMARICUS FORMOSUS. Thor.

Syn. : 1887. *Camaricus formosus*, Thor. (Viaggio di L. Fea, Cet. 2), Primo Saggio sui Ragni Birmani, loc. cit., p. 262.

1890. „ *fornicatus*, Thor., Arachnidi di Nias e di Sumatra, p. 60.

Description of Plate 4 ♂—*a*, figure of spider magnified (abdomen too short in drawing); *b*, natural size; *c*, eyes and falces; *d*, profile of spider; *e*, left palpus

Sternum, falces, labium, and bases of the legs yellow. Dark brown coloured marking on lower side of abdomen with a light yellow margin. This species much resembles *Tharpyna diademata*, L. Koch, Arachniden Australis, page 548, Taf. 42, figs. 2—3, but it differs remarkably from this species in the position of middle front eyes, which are nearer to one another than each one is to the corresponding side eye. Found in Burma, Sumatra, and Penang. Specimen in my collection. ♀ unknown.

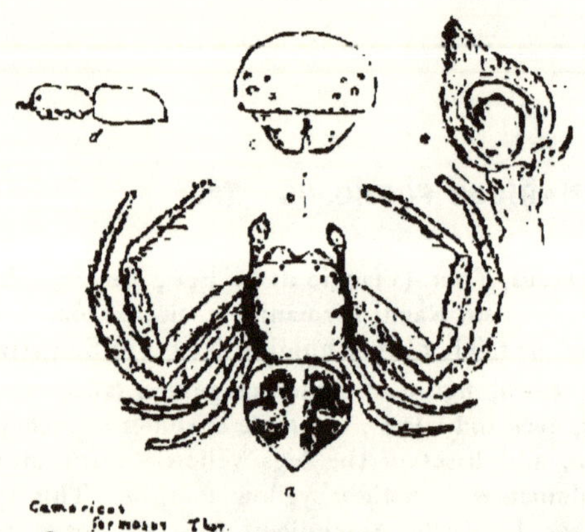
d
a
Camaricus
formosus Thor.
Penang

EPEIRA BECCARII. Thor.

Syn.: 1878. *Epeira beccarii*, Thor., Studi sui Ragni Malesi e Papuani II., Ragni di Amboina, cet., pp. 65—297.

1881. „ „ id., ibid. III., Ragni dell'Austro-Malesi, cet., p. 119.

Description of Plate 5 ♀ —*a*, figure of spider magnified; *b*, natural size; *c*, eyes; *d*, epigyne; 5*a*, snare

This spider I found in considerable numbers on a clump of mangroves growing at the side of the Deli road, about two miles from Singapore. Their webs are most beautifully constructed. The circular snare is placed horizontally with the edges somewhat turned up, and the centre raised in a cornicopea like shape, the upper part being slightly turned over; half way up this tube the egg-cocoon is placed.

Below the snare is a network, larger than the snare, formed of irregular lines crossing and recrossing to protect the mouth of the tube, &c., on the underside. The snare is eight inches in diameter, and the tube three inches long. The tube was braced to the surrounding twigs and leaves to keep it in an upright position. The wonderful regularity of the circular snare, and the beautiful curve of the ascending tube, together with its perfect adaption as a means of supplying food and protection to its constructor and her progeny, made it one of the most interesting objects I have ever seen. Hentz describes a somewhat similar web as made by *Epeira labyrinthea*, Hentz. See Hentz's Spiders of the United States, page 111, pl. 13, fig. 3, pl. 19, figs. 124 and 133. Found in Amboina and Singapore, but not common in Singapore. ♂ unknown.

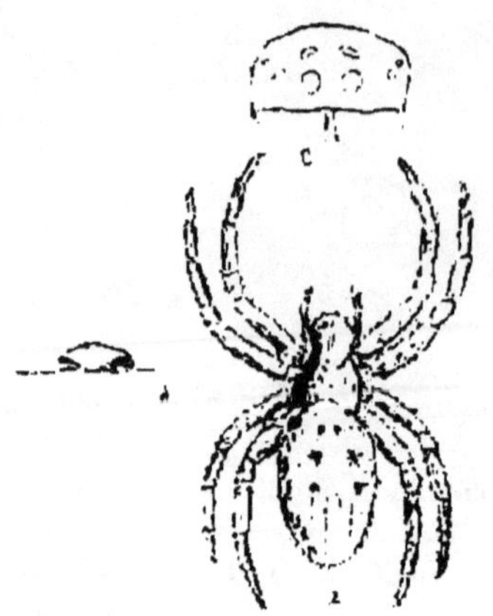

Epeira beccarii
Thorell
Sinoanire

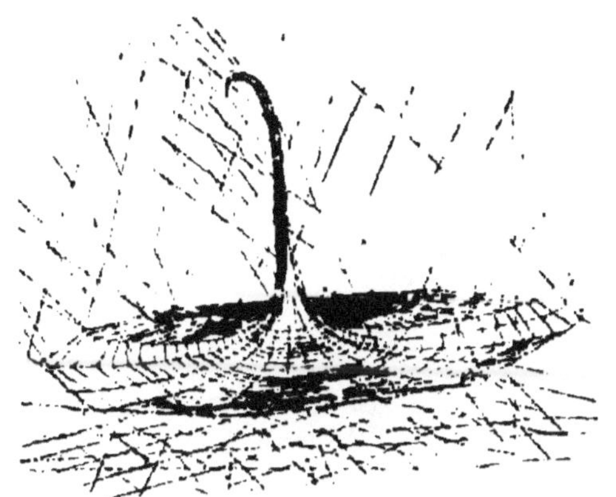

EUOPHRYS PYGAEA. Thor.

Syn. : 1891. *Euophrys pygaea*, Thor., Spindlar fran Nikobarerna och andra delar of Södra Asien. (K. Svenska Vetenskaps—Akademiens Handlingar, XXIV., No. 2), p. 135.

Description of Plate 6 ♀—*a*, figure of spider magnified; *b*, natural size; *c*, eyes; *d*, epigyne; *e*, labium and maxillae; *f*, spider in profile.

Length of legs, i.—3, ii.—quite 3, iii.—nearly 3, iv.—3½ milemetres.

Cephalothorax and abdomen sparsely covered with white hairs which do not show until the specimen is taken out of spirits and dried. Found by me in Penang. Specimen in my collection. ♂ unknown.

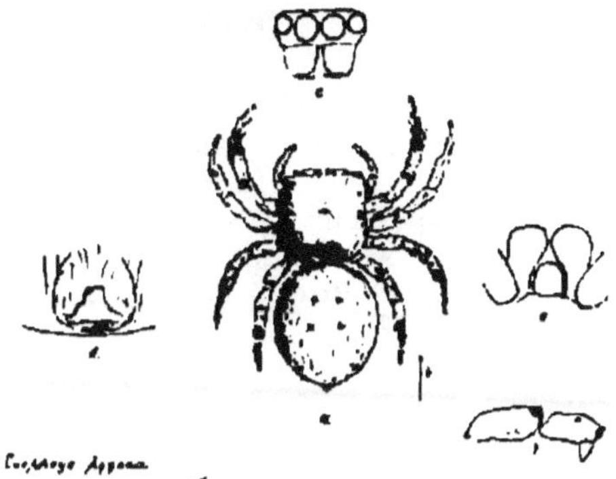

Thor.

Penang

GASTERACANTHA DORIAE. Sim.

Syn.: 1877. *Gasteracantha doriae*, Sim., Etudes Arach. X., Arach. Nouv. ou peu connus, in Ann. de la Soc. Ent. de France, 2ᵉ Ser., VII., p. 232, pl. 3, fig. 3.

1879. " *harpax*, Camb., on some new and little known Spec. of Aran., with remarks on the gen. Gasteracantha, in Proceed. of the Zool. Soc. of London, 1879, p. 284, pl. XXVI, fig. 9.

1889-90. " *doriae*, Thor., Studi cet. IV., Vol. 1, p. 57.

Description of Plate 7 ♀—*a*, figure of spider natural size; *b*, eyes; *c*, centre of web showing beginning of outer spiral.

Snare seventeen inches diameter, constructed in a coffee-tree in Mr. Davidson's plantation on the mainland near Singapore. It had thirty rays which did not meet in the centre, but were joined by a circular thread. Outer spiral commences two inches out from this central ring and makes twenty-six revolutions, The spider was sitting in the centre of the web, head downwards. It has also been found at Sarawak in Borneo. ♂ unknown.

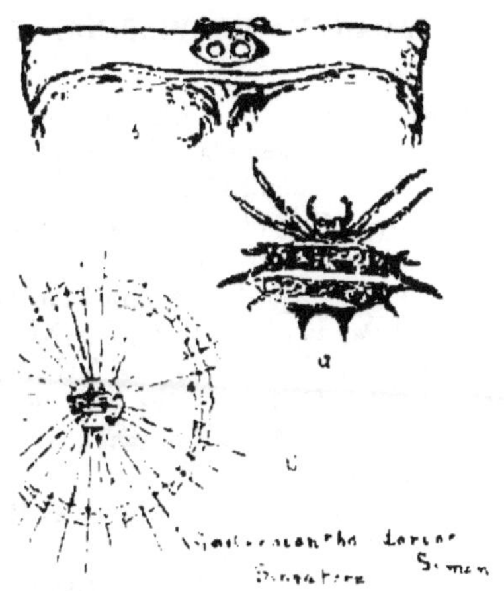

Gasteracantha larvae Simon
Singapore

GASTERACANTHA BREVISPINA. Dol.

Syn.: 1837. *Gasteracantha cuspidata*, C. L. Koch, Die Arach. IV., p. 22 (ad partem: exempla "tuberculo-cephalothoracis singulo, conico").

1857. *Plectana brevispina*, Dol., Bijdrage tot de Kennis der Arachniden van de Indischen Archipel. p. 423.

1859. „ *flavida*, id., Tweede Bijdrage ibid., p. 43, tab. XIII., fig. 3.

1859. „ *roseo-limbata*, id, ibid., p. 43, tab. XIII., fig. 1

1859 „ *mediofusca*, id., ibid., p. 44, tab. XIII., fig. 9.

1859. *Gasteracantha mammeata*, Thor., Nya exot. Epeir, in Ofvers. af Vet.-Akad. Forhandl. XVI., p. 301.

? 1859. „ *guttata*, id., ibid.

1868. „ *mammeata*, id., Freg. Eugenies Resa, Arach., I., p. 18.

? 1868. „ *guttata*, id., ibid., p. 19.

1871. „ *suminata*, L. Koch. Die Arach. Austral., p. 11, Tab. I., fig. 7—7a.

1878. „ *brevispina*, Thor. Studi cet., II. Ragni Malesi e Papuani raccolti dal Prof. O. Beccari, pp. 17 to 294.

? 1879. „ *observatrix*, Camb., On some new and little known spec. of Aran., with remarks on the gen. Gasteracantha, in Proceed. of the Zool. Soc Lond., 1879, p. 291, Pl. XXVII., fig. 21.

1881. „ *brevispina*, Thor., Studi cet., III., Ragni dell'Austro-Malesia, cet., p. 27.

1882. „ *flavida*, Van Hass., Midden Sumatra, cet., Aran., p. 15.

? 1882. „ *mammosa*, id., ibid.

1889-90. „ *brevispina*, Thor., Studi cet. IV., Vol. I., p. 63.

Description of Plate 8.—*a*, figure of spider magnified; *b*, natural size; *c*, eyes; *d*, palpus of ♂.

This spider makes a similar web to that of *Gasteracantha doriae* Sim but only twelve inches in diameter, and also sits in the centre of web, head downwards. It was found at Mr. Davidson's plantation in a coffee-tree. Common in the Malayan Archipelago.

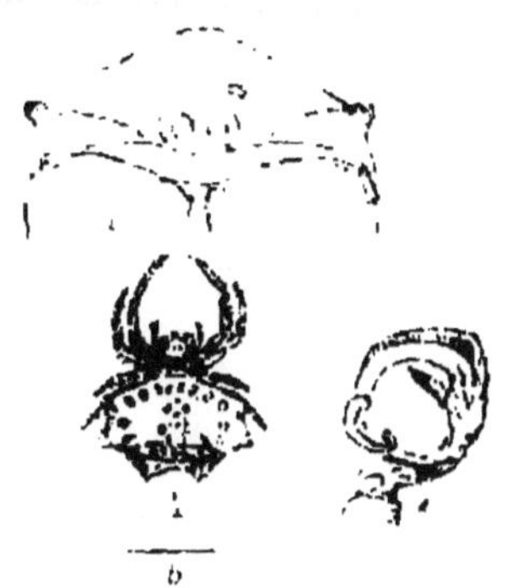

Gasteracantha brevispina
Dol

MANTIUS RUSSATUS. Thor.

Syn. : 1891. *Mantius russatus*, Thor., Spindlar fran Nikobarerna och andra delar af Södra Asien. (K. Svenska Vatenskaps—Akademiens Handlingar, XXIV., No. 2), p. 140.

Description of Plate 9 ♀—*a*, figure of spider magnified ; *b*, natural size ; *c*, labium and maxillae ; *d*, eyes ; *e*, epigyne.

Total length, $6\frac{3}{4}$; length of cephalothorax, $2\frac{5}{8}$; breadth, 2, in front $1\frac{3}{4}$; length of abdomen, $3\frac{3}{4}$; breadth, $2\frac{1}{2}$ millim. Length of legs, i.—$4\frac{1}{2}$, ii.—$4\frac{1}{3}$, iii.—$5\frac{1}{4}$, iv.—$5\frac{1}{2}$ millim. Length of patella + tibia, i.—about $1\frac{3}{4}$, patella+tibia, iii.—$1\frac{5}{8}$, patella + tibia, iv.—$1\frac{3}{4}$, metatarsus + tarsus, iv.—$1\frac{1}{8}$ millim (Thorell).

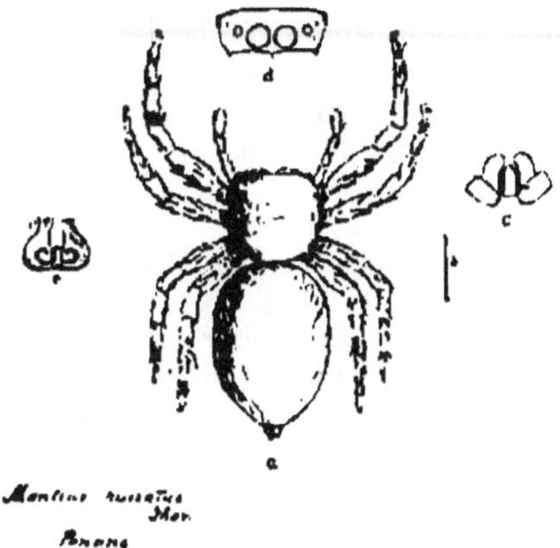

Mantius russatus Thor.
Penang

CHRYSILLA VERSICOLOR. C. L. Koch.

Syn.: 1846. *Plexippus versicolor*, C. L. Koch, Die Arach. xiii., p. 103, tab. CCCCXLIX., fig. 1165 (= ♂ ad.).

1848. *Maevia picta* id., ibid., xiv., p. 72, tab. CCCCLXXVIII., fig. 1328 (= ♂ jun.).

1891. *Chrysilla versicolor*, Thor. Spindlar fran Nikobarerna och andra delar af Södra Asien. (K Svenska Vatenskaps—Akademiens Handlingar XXIV., No. 2), p. 117.

Description of Plate 10 ♂.—*a*, figure of spider magnified; *b*, natural size; *c*, spider in profile; *d*, eyes and falces; *e*, maxillae and labium; *f*, right palpus in front; *g*, do. in profile; *h*, epigyne of ♀.

♂ Total length, 4½; cephalothorax, 2⅓, breadth about 1¾, in front at least 1½; abdomen, 2½, breadth about 1¼ millim. Legs, i.—almost 5, ii.—about 4⅗, iii.—almost 5, iv.—5½ millim. in length; patella + tibia, iii.—almost 1⅔, patella + tibia, iv.—1¾; metatarsus + tarsus, iv.—1¾ millim.

♀ Total length, 5¼; cephalothorax, 2¼ (equal to patella + tibia and ⅗ metatarsus iv.), breadth a little more than 1½, in front 1⅓; abdomen 3, breadth 1⅓ millim. Legs, i.—a little more than 4, ii.—3¼, iii.—4¾, iv.—5⅓ millim.; patella + tibia, i.—1½, patella + tibia, iii.—1⅓, patella + tibia, iv.—a little more than 1½, metatarsus + tarsus, iv.—almost 1⅗ millim. (Thorell). Abdomen and part of cephalothorax covered with scale like white hairs.

Koch's ♂ and ♀ found at Bintang. Both sexes found by me at Penang. ♂ by Dr. Klein at Padang Sumatra. Found also in Singapore.

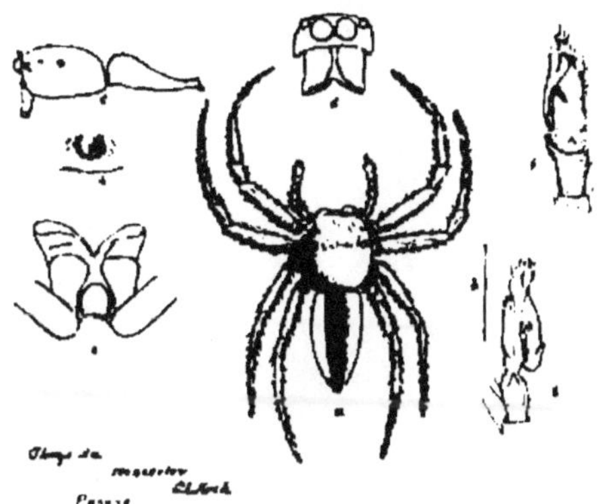

MAEVIA ALTERNANS. C. L. Koch.

Syn. : ? 1846. *Plexippus vitatus*, C. L. Koch, Die Arach., xiii., p. 125, tab. CCCCLIII., fig. 1185 (= ♀ jun.).

1846. *Hyllus alternans*, id., ibid., p. 160, tab. CCCCLX., fig 1222.

1891. *Maevia alternans*, Thor., Spindlar fran Nikobarerna och andra delar af Södra Asien. (K. Svenska Vatenskaps—Akademiens Handlingar, XXIV., No. 2), p. 122.

Description of Plate 11 ♂—*a*, figure of spider magnified ; *b*, natural size ; *c*, maxillae and labium ; *d*, spider in profile ; *e*, falce from in front ; *f*, right palpus in front ; *g*, do. side view ; *h*, eyes and falces.

Total length, ♂ $4\frac{1}{2}$; cephalothorax, $2\frac{1}{3}$, breadth at least $1\frac{5}{8}$, breadth in front almost $1\frac{1}{3}$; abdomen, $2\frac{1}{6}$, breadth $1\frac{1}{3}$ millim. Legs, i.—6, ii.—almost 5, iii.—$5\frac{1}{4}$, iv.—6 millim. long ; patella + tibia, i.—about $2\frac{1}{3}$, patella + tibia, iii.—a little more than $1\frac{1}{2}$, patella + tibia, iv.—at least $1\frac{5}{8}$, metatarsus + tarsus, iv.—$1\frac{5}{8}$ millim. (Thorell).

White and gold metallic scales on cephalothorax and abdomen. Koch's ♀ found at Bintang, and the ♂ at Puloloz, East India. ♂ found by me at Penang.

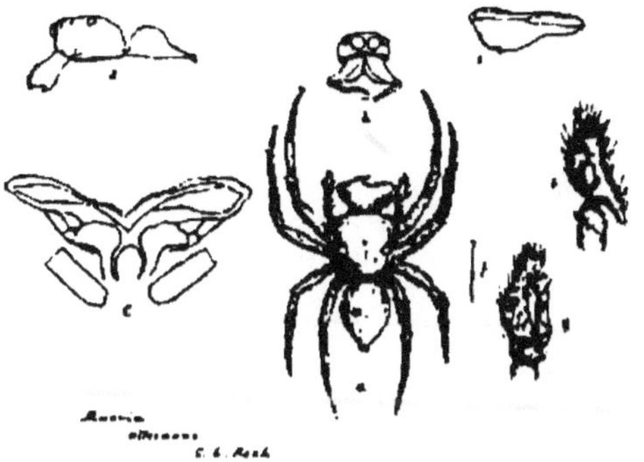

EPHIPPUS D'URVILLEI. Walck.

Syn.: 1837. *Attus d'urvillei*, Walck, H. N. d. Ins. Apt., I., p. 459.
1881. *Ephippus d'urvillei*, Thor. Studi sui Ragni Malesi e Papuani III., Ragni dell'Austro-Malesi, etc., p. 653.

Description of Plate 12 ♂.—*a*, figure of spider magnified; *b*, natural size; *c*, spider in profile; *d*, eyes and falces; *e*, maxillae and labium; *f*, left palpus from outside; *g*, do. in front.

♂ Total length, 11¾; cephalothorax, 6¼, breadth almost 5⅛, in front 3½; abdomen, 6, breadth 4 millim. Legs, i.—16, ii.—12, iii.—15, iv.—12⅓; patella + tibia, iii.—4⅓; patella + tibia, iv.—3½; metatarsus + tarsus, iv.—4¼ millim. (Thorell.)

♀ Total length 10⅓; cephalothorax, 5½, breadth almost 4⅛, in front 3⅓; abdomen, 5½, breadth 4 millim. Legs, i.—10⅔, ii.—9, iii.—12¼, iv.—11⅓; patella + tibia, iii.—4¼, patella + tibia, iv.—3¾; metatarsus + tarsus, iv.—3½ millim. (Thorell).

Coloured scales on cephalothorax and abdomen of ♂.

Found in New Guinea, Arau Islands, and Northern Australia.

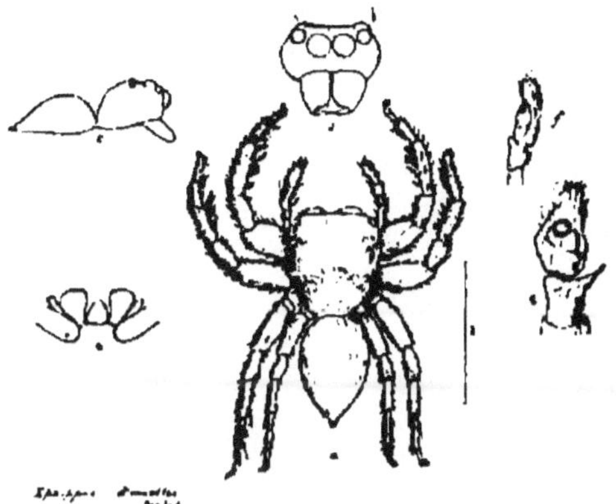

MEGATIMUS SEVERUS. Thor.

Syn. : 1891. *Megatimus severus*, Thor., Spindlar fran Nikobarerna och andra delar af Södra Asien. (K. Svenska Vatenskaps—Akademiens Handlingar, XXIV., No. 2), p. 129.

Description of Plate 13 ♀—*a*, figure of spider magnified ; *b*, natural size ; *c*, spider in profile ; *d*, eyes and falces ; *e*, maxillae and labium ; *f*, epigyne.

Total length, $11\frac{1}{3}$; cephalothorax, almost 6, breadth $4\frac{1}{3}$, in front 3 ; abdomen, $6\frac{1}{2}$, breadth, $4\frac{1}{2}$ millim. Legs, i.—$11\frac{1}{3}$, ii.—$9\frac{1}{3}$, iii.—$9\frac{1}{2}$, iv.—$12\frac{3}{4}$ millim. ; patella + tibia, i.—a little more than 5, patella + tibia, iii.—$3\frac{1}{4}$, patella + tibia, iv.—a little more than 4, metatarsus + tarsus, iv.—$3\frac{1}{2}$ millim. (Thorell).

Coloured metallic scales on front of falces. One specimen ♂ found by me in Penang. ♀ unknown. Type of the *Genus megatimus*. Thor.

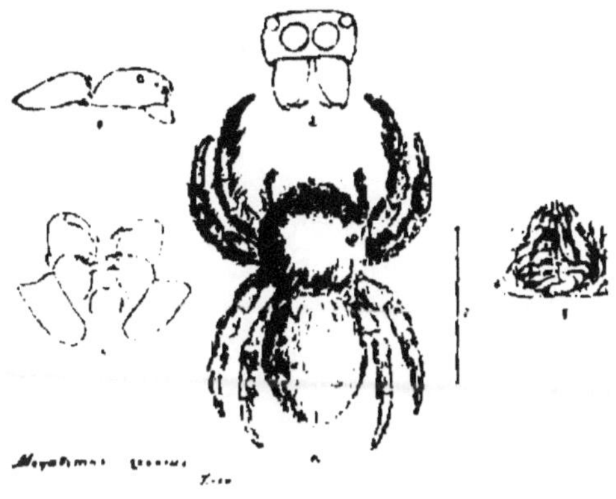

CARRHOTUS VIDUUS. C. L. Koch.

Syn.: 1846. *Plexippus viduus*, C. L. Koch, Die Arachn. xiii., p. 104, tab. CCCCXLIX., fig. 1166.

? 1846. „ *albo-lineatus*, id., ibid., p. 105, tab. CCCCXLIX., fig. 1167.

1891. *Carrhotus viduus*, Thor., Spindlar fran Nikobarerna och andra delar af Södra Asien. (K. Svenska Vatenskaps—Akademiens Handlingar, XXIV., No. 2), p. 142.

Description of Plate 14 ♂.—*a*, figure of spider magnified; *b*, natural size; *c*, spider in profile; *d*, eyes and falces; *e*, maxillae and labium; *f*, left palpus in front.

Total length, 5¼; cephalothorax, 2½, breadth 2, about 1¾ in front; abdomen, 3, breadth 1¾ millim.; legs, i.—5, ii.—4½, iii.—5, iv.—4¼ millim.

A band of white hairs down each side of abdomen. White hairs on sides of cephalothorax.

Koch's ♂ found in Bintang, and *P. albolineatus*, C. L. Koch, which Professor Thorell believes to be the ♀ of *Carrhotus viduus*, was found in Java. ♂ found by me in Penang.

Carrhotes
Penang

HASARIUS SANNIO. Thor.

Syn.: 1877. *Plexippus (?) sannio*, Thor., Studi sui Ragni Malesi e Papuani I., Ragni di Selebes, p. 617 (277).

1891-92. *Hasarius sannio*, id., ibid. IV., Ragni dell' Indo-Malesia, Vol. 2, p. 429.

Description of Plate 15 ♂—*a*, figure of spider magnified; *b*, natural size; *c*, spider in profile; *d*, eyes and falces; *e*, maxillae and labium; *f*, right palpus from inside; *g*, do. in front.

Total length, 5; cephalothorax, $2\frac{5}{8}$, breadth $1\frac{3}{4}$, in front $1\frac{2}{3}$ millim.; abdomen, $2\frac{1}{3}$, breadth $1\frac{1}{2}$ millim. Legs, i.—$5\frac{1}{2}$, ii.—$4\frac{1}{3}$, iii.—(5), iv.—$4\frac{2}{3}$; patella + tibia, iv.—almost $1\frac{1}{3}$, metatarsus + tarsus, iv.—$1\frac{1}{2}$ millim. (Thorell).

Cephalothorax and abdomen densly covered with long hairs, also the legs, on which are also spines. ♂ found in Celebes and Kandari, and by me in Penang. ♀ unknown.

Van Hasselt describes (Midden Sumatra, cet., Aran., p. 48, Pl. V., fig. 15) a spider which he calls *Plexippus sannio* ♀ jun., but which E. Simon (Arach. recueillis par M. Weyers à Sumatra, cet., p. 4) considers to be a different species, and calls it *Ergane coronata*.

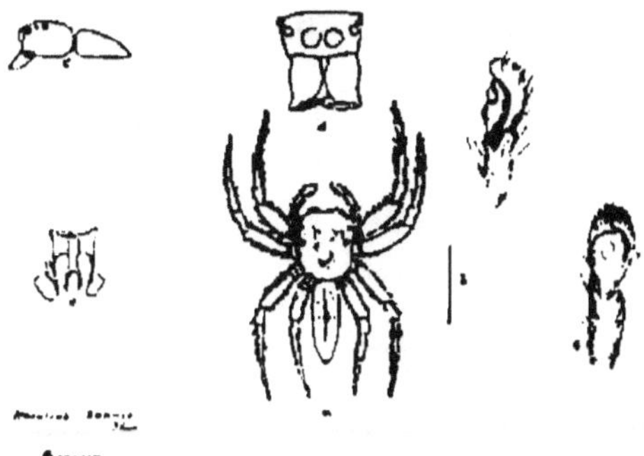

HASARIUS VIRENS. Thor.

Syn.: 1890. *Hasarius virens*, Thor., Arachnidi di Nias e di Sumatra, p. 89.
1891. „ „ id., Spindlar fr. Nikobarerna ach andra delar af Södra Asien. (K. Svenska Vatenskaps—Akademiens Handlingar, XXIV., No. 2), p. 147.

Description of Plate 16 ♀.—*a*, figure of spider magnified; *b*, natural size; *c*, spider in profile; *d*, eyes and falces; *e*, maxillae and labium; *f*, epigyne.

Total length, 5; cephalothorax, almost $2\frac{1}{3}$, breadth about $1\frac{3}{4}$, in front a little more than $1\frac{1}{2}$; abdomen, $2\frac{3}{4}$, breadth almost 2 millim. Legs, i.—$4\frac{1}{3}$, ii.—$4\frac{1}{4}$, iii.—$4\frac{1}{2}$, iv.—5 millim.; patella + tibia, i.—$1\frac{1}{3}$, patella + tibia, iii.—almost $1\frac{1}{2}$, patella + tibia, iv.—$1\frac{1}{2}$; metatarsus + tarsus, iv.—almost $1\frac{1}{2}$ millim. (Thorell.)

♀ found in Sumatra, and by me in Penang. ♂ unknown.

Has considerable resemblance to *Mantius russatus*, Thor., but differs from it by the front middle eyes being larger and projecting more; also by the lighter coloured abdomen, which has a band of dark hair around the front. Similar dark hairs on the femoral joints of legs.

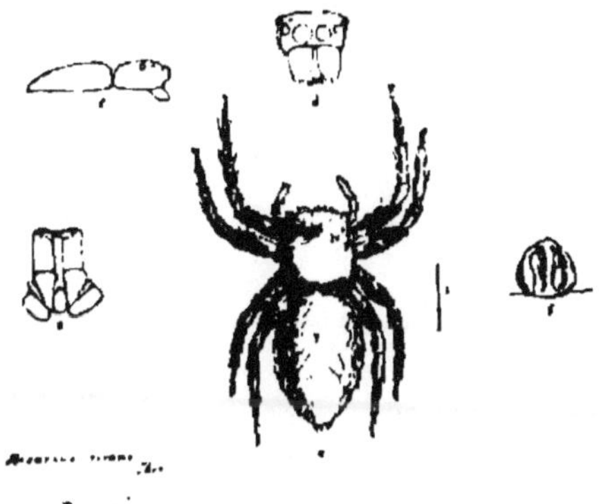

ARGYRODES SUMATRANUS. Thor.

Syn.: 1889-90. ***Argyrodes sumatranus***, Thor. Studi sui Ragni Malesi, etc., Part IV., Vol. 1., p. 247.

Description of Plate 17 ♀.—*a*, figure of spider magnified; *b*, natural size; *c*, spider in profile; *d*, epigyne; *e*, eyes and falces; *f*, cephalothorax in profile and palpus of ♂; *g*, left palpus in front.

♂ Total length, 3½; cephalothorax, about 1¾, breadth, about 1; breadth of clypeus, about ⅛; abdomen, 2⅓, breadth, a little more than 1, height, 1⅜ millim. Legs, i.—(13½), ii.—7½, iii.—4½, iv.—7½; patella+tibia, iv.—2 millim. (Thorell.)

♀ Total length, 4⅔; cephalothorax, about 1⅗, breadth, about 1; breadth of clypeus, almost ½; abdomen, 2⅔, breadth, 2½, height, almost 3½ millim. Legs, i.—12, ii.—7, iii.—4⅔, iv.—6½; patella+tibia, iv.—2 millim. (Thorell.)

This spider has a very close resemblance to ***Theridion miniaceum***, Dol., and ***Theridion sundiacum***, Dol. Found in Sumatra and Singapore.

ARGYRODES SUMATRANUS. Thor.

Syn.: 1889-90. *Argyrodes sumatranus*, Thor. Studi sui Ragni Malesi, etc., Part IV., Vol. 1., p. 247.

Description of Plate 17 ♀.—*a*, figure of spider magnified; *b*, natural size; *c*, spider in profile; *d*, epigyne; *e*, eyes and falces; *f*, cephalothorax in profile and palpus of ♂; *g*, left palpus in front.

♂ Total length, 3½; cephalothorax, about 1¾, breadth, about 1; breadth of clypeus, about ⅓; abdomen, 2⅛, breadth, a little more than 1, height, 1⅜ millim. Legs, i.—(13½), ii.—7½, iii.—4½, iv.—7½; patella+tibia, iv.—2 millim. (Thorell.)

♀ Total length, 4¾; cephalothorax, about 1⅗, breadth, about 1; breadth of clypeus, almost ½; abdomen, 2⅛, breadth, 2½, height, almost 3½ millim. Legs, i.—12, ii.—7, iii.—4⅗, iv.—6½; patella+tibia, iv.—2 millim. (Thorell.)

This spider has a very close resemblance to ***Theridion miniaceum***, Dol., and ***Theridion sundiacum***, Dol. Found in Sumatra and Singapore.

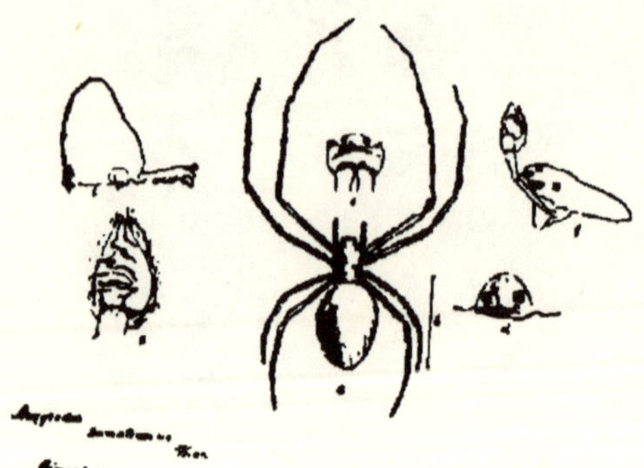

ULOBORUS QUADRI-TUBERCULATUS. Thor.

Syn.: 1892. *Uloborus quadri-tuberculatus*, Thor. In MSS.

Description of Plate 18 ♀.—*a*, figure of spider magnified; *b*, natural size; *c*, spider in profile; *d*, maxillae and labium; *e*, epigyne; *f*, do. in profile; *h*, cocoon magnified; *i*, do. natural size. 18*a*. Web of spider in Pine Apple plant reduced.

This spider is not uncommon in Singapore. It makes a curious snare in the centre of the Pine Apple plant (*Bromelia ananas*), which it forms peculiarly to suit the shape of the plant, and I have not found its snare elsewhere. The Pine Apple is a South American plant, and has only of late years been introduced into this part of the world. Was the spider introduced with it, or is it a native which has adapted its snare to the introduced plant? The snare is horizontal, loosely made of rays on the top, in the centre of which on the lower side the spider sits. 24 top rays not adhesive. About 24 lower or side rays on which there is a 14 thread spiral. The snare is held open at the bottom so that the spider may drop through. When disturbed the spider hangs on by a ½ inch thread from centre of web and swings itself violently to and fro. The little spiked cocoon is attached to the top rays, close to where the spider sits in the centre.

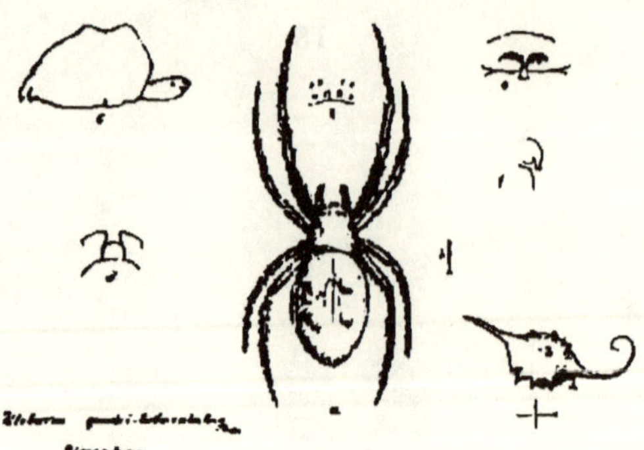

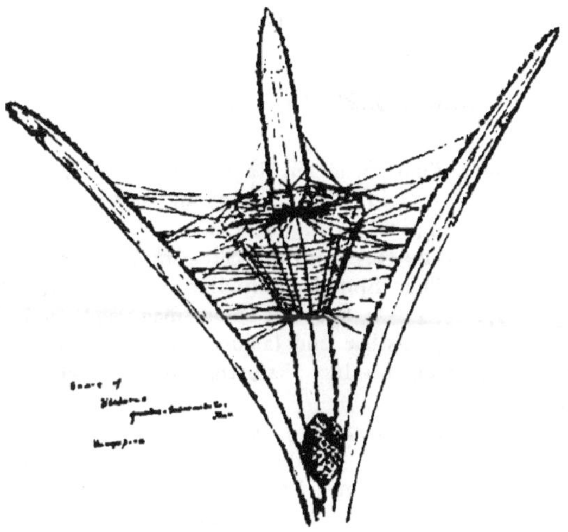

ARGYROEPEIRA STRIATA. Thor.

Syn.: 1877. *Meta striata*, Thor. Studi sui Ragni Malesi, etc., Vol. 1., p. 427 (87).
1878. „ „ id., ibid., Vol. 2, pp. 97 and 297.
1881. „ „ id., ibid., Vol. 3, p. 130.
1882. „ „ Van Hass., Midden Sumatra, etc., Araneae, p. 25.
1891. „ „ Thor. Spindlar fran Nikobarerna, etc., p. 47.

Description of Plate 19 ♀.—*a*, figure of spider magnified; *b*, spider in profile; *c*, natural size; *d*, epigyne; *e*, maxillae and labium; *f*, eyes and falces; *g*, right palpus of ♂ from outside; *h*, do. from inside. Plate 19*a*.—*a*, snare with trap line; *b*, snare without trap line.

Total length, 5½; cephalothorax, 2⅓, breadth nearly 1¾, in front nearly 1; abdomen, 3⅓, breadth, 2¼, height, 3 millim. Legs, i.—20, ii.—12, iii.—5⅝, iv.—10 millim.; patella + tibia, iv.—3 millim.; Falce nearly 1⅔ millim. long (Thorell).

The following are my notes on the snares of this species:—

No.	Inches diameter.	Angle.	Rays.	Inner spiral.	Free zone inner to outer spiral.	outer spi.
1	9	45°	18	6	2	20
2	10	45°	24	6	1½	20
3	—	Horizontal.	28	4	1	28
4	16	do.	33	0	1½	28
5	12	45°	20	5	1	19
6	—	Horizontal.	24	3	1½	27
7	—	45°	18	3	1	(28 / 26)
8	12	45°	21	4	1½	36

In Nos. 1, 2, and 8, Spider was sitting in centre of web, head downwards; in No. 6 on the under side; and in Nos. 4, 5, and 7 it had a nest under a leaf; and in 5 and 7 there was an open segment in the snare and a trap line to the nest. However, in No. 7, I believe the open segment must have been caused by an accident, as there were 28 rays on the upper side and only 26 on the lower, and some of the spiral threads were carried partially across open segment up to a leaf; also, Spider was in centre of web when I first saw it.

The rays mostly meet in a central ring, ¼—½ inch in diameter, but in two snares, Nos. 5 and 8, they met in a reticulation. The snares were made near the ground in low bushes. The Spider in No. 8 snare when disturbed stretched out its legs after the manner of *Tetragnatha extensa* Linn. I also noticed that this species of Spider has the power of darkening down its brilliant colouring when frightened. The colour returning when it is dead.

When living it is of a brilliant golden-green colour, and is found in all parts of the Malaysian Archipelago.

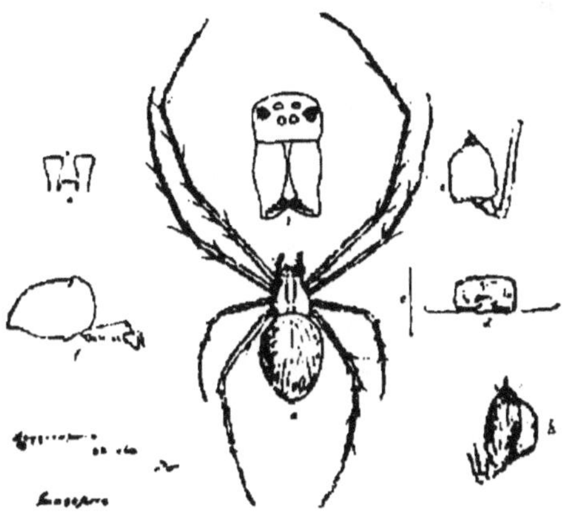

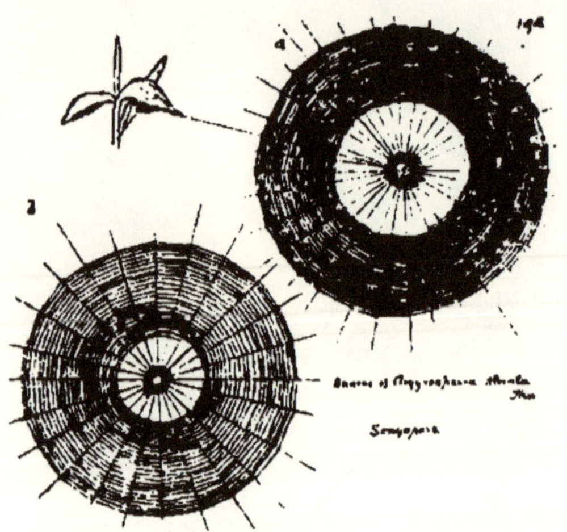

EPEIRA UNICOLOR. Dol.

Syn. : 1857. *Epeira unicolor*, Dol., Bijdrage tot de Kennis der Arachniden, etc., p. 419.
1859. „ „ id., Tweede Bijdrage, etc., tab. XI., fig. 1.
1878. „ „ Thor., Studi sui Ragni Malesi, etc., Part II., pp. 53 and 296.
1881. „ „ id., ibid., Studi, etc., Part III., p. 96.

Description of Plate 20 ♀.—*a*, figure of spider magnified; *b*, natural size; *c*, spider in profile ; *d*, eyes. Plate 20*a*.—Snare.

Total length, 23½ ; cephalothorax, 10¾, breadth 8½, in front 3¾ ; abdomen, 15½, breadth 14½ millim. Legs, i.—31½, ii.—29½, iii.—19, iv.—28 ; patella + tibia, iv.—8¾ millim. Falce, 3¾ millim long. (Thorell.)

This spider makes a large horizontal circular snare on low shrubs, over which it suspends leaves, under one of which it conceals itself. Doleschall also describes how at Amboina he found it between rolled up leaves. Found by me in Singapore.

♂ unknown.

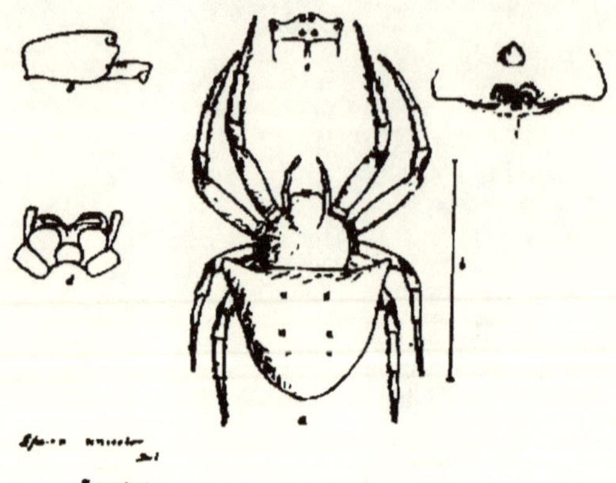

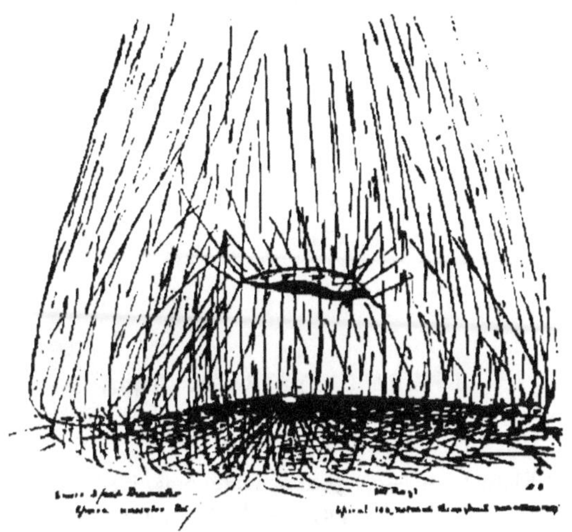

EPEIRA CALYPTRATA. Thor.

Syn.: 1892. *Epeira calyptrata*, Thor. in MSS.

Description of Plate 21 ♀.—*a*, figure of spider magnified; *b*, natural size; *c*, spider in profile; *d*, maxillae and labium; *e*, eyes and falces; *f*, epigyne. 21*a*.—Snare.

Of this species I captured in Singapore three in their snares. No. 1 snare was circular, 4 inches in diameter, with ray line to nest under twisted bamboo leaf. No free segment. 18 rays meeting in irregular ring. Inner spiral 3. Free zone inner to outer spiral ⅝ inch.* Outer spiral 10 at trap ray, 18 opposite. Trap ray, one of the ordinary rays, was in the upper part, and to one side. No. 2—Perpendicular snare 8 inches across, consisting of 40 rays meeting in ⅛ inch ring. Inner spiral 4. Free zone 1 inch. Outer spiral 16 at trap ray, 30 opposite. Trap ray attached to nest in upper quarter, and to one side. No. 3—Perpendicular snare 4 × 8 inches with trap ray to nest under twisted bamboo leaf Rays 30 meeting in ¼ inch ring. Inner spiral 4. Free zone 1¼ inch. Outer spiral at top 16, at bottom, 37 rays. In the nest of No. 3 there were two cocoons made of loosely-twisted brownish silk, and each containing about 20 newly hatched spiders.

♂ unknown.

* Measurement of free zone in all cases is space between inner and outer spiral.

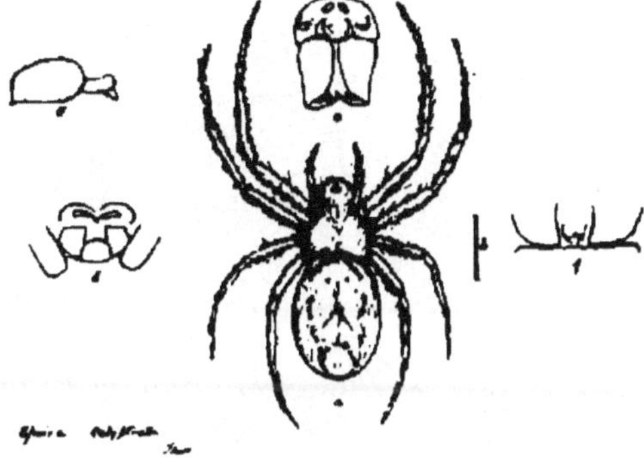

Epeira [illegible] [illegible]

Singapore

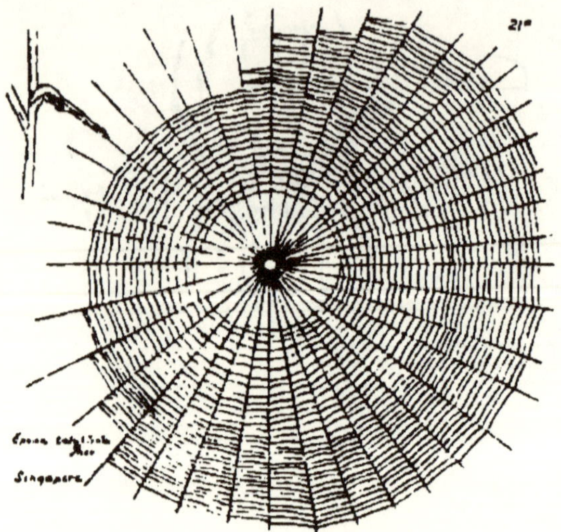
21a
Singapore

CALLINETHIS ELEGANS. Thor.

Syn.: 1877. *Meta elegans*, Thor., Studi sui Ragni Malesi, etc., Part I., p. 410 (76).
1889-90. *Callinethis elegans*, id., ibid., Studi, etc., Part IV., Vol. I., p. 193.

Description of Plate 22 ♀.—*a*, figure of spider magnified; *b*, natural size; *c*, eyes; *d*, spider in profile; *e*, epigyne. 22*a*—Snare.

Total length, 9; cephalothorax 3½, breadth 2⅘, in front 1½; abdomen 6⅓, breadth 3½ millim. Legs, i.—17⅓, ii.—15, iii.—8⅓, iv.—13½; patella + tibia, iv.—4 millim; falce, 2 millim. long (Thorell.)

I only obtained one snare of this species in Singapore, about 5 feet from the ground. It was a beautiful, almost perfect, perpendicular circle 16 inches in diameter, and the spider was sitting in the centre head downwards. Rays 30 (not adhesive) meeting in ½ inch ring. Inner spiral 4. Free zone 1½ inch. Outer spiral 45, adhesive.

This species has also been found in the Celebes. ♂ unknown.

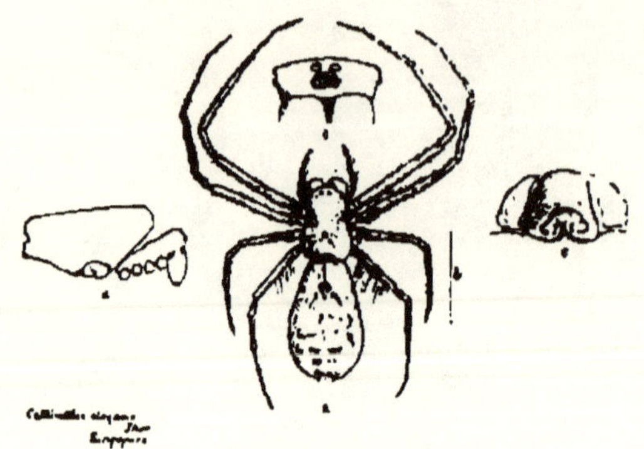

GEA DECORATA. Thor.

Syn.: 1889-90. *Gea decorata*, Thor., Studi sui Ragni Malesi, etc., Part IV., Vol. 1, p. 104.

Description of Plate 23 ♀.—*a*, figure of spider magnified; *b*, natural size; *c*, spider in profile; *d*, maxillae and labium; *e*, eyes and falces; *f*, epigyne; *g*, epigyne greatly magnified. 23*a*.—snare.

Plate 24 ♂.—*a*, figure of spider magnified; *b*, natural size; *c*, spider in profile; *d*, eyes and falces; *e*, maxillae and labium; *f*, left palpus from inside; *g*, right palpus and eyes from above.

Total length, 6; cephalothorax, almost 3, breadth, 2½, in front a little more than 1; abdomen, 3⅜, breadth, 2⅜ millim. Legs, i.—10¾, ii.—10½, iii.—6¼, iv.—almost 10½ millim.; patella+tibia, iv.—almost 3¼ millim. (Thorell.)

I obtained three snares of this species in Singapore; two of them among grass and ferns, and all perpendicular. Spider sitting in centre, head downwards.

No. 1, 8 inches in diameter; rays, 48, meeting in a reticulation; inner spiral, 15, not adhesive, and getting wider as it goes outward until it joins the outer spiral, so there is no free zone; outer spiral, adhesive, 28 at top and 40 at bottom; small protecting reticulation behind the web at the Spider's back.

No. 2, 5 inches in diameter; 56 rays meeting in a ring; inner spiral running through free zone in a gradually enlarging spiral; outer spiral, top 24, bottom 50.

In this web there was a ♂ Spider, but no ♀; and it is from this Spider that the ♂ drawing is made. Though the Spider was not in the web but on one of the rays I do not think I am wrong in describing it as the ♂ of this species, both from the style of the web and the position in which I found it.

No. 3, 7 inches in diameter; rays, 56; inner spiral, 16, spreading as it goes outward; free zone, 1 inch diameter; across free zone, 3½ inches; outer spiral, 39.

This species is also found in Sumatra.

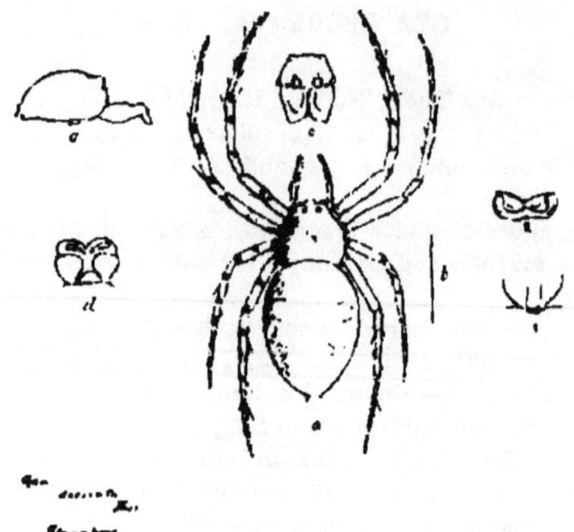

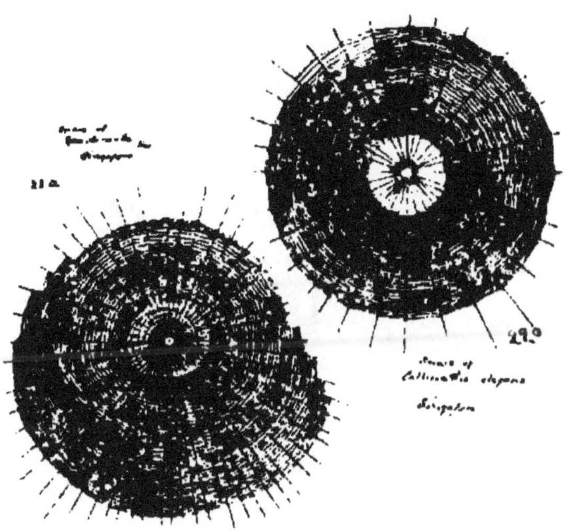

21a
29.

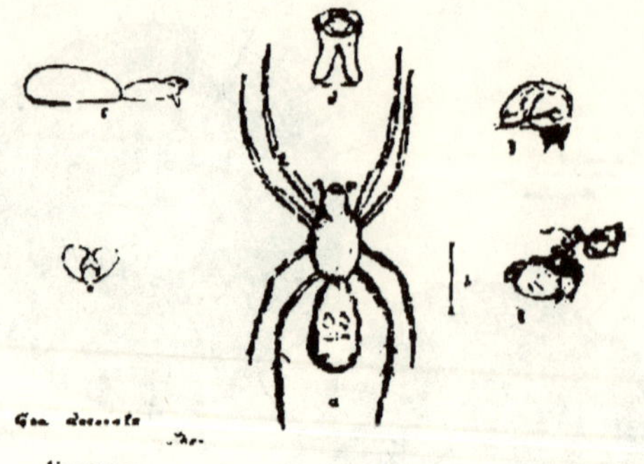

Gea decorata

Singapore

CAEROSTRIS PARADOXA. Dol.

1859. *Epeira paradoxa*, Dol., Tweede Bijdr., etc., p. 37, pl. IX., fig. 11—11c; pl. X., figs. 8—8c.

? 1880. *Caerostris mitralis*, Thor., Freg. Eugenies resa., Arachn., 1, 1, 4.

1880. „ *paradoxa*, Butl., Proceed. of the Zool. Soc. of London, 1879, p. 732, Pl. LVIII., figs. 5—5b.

1890. „ „ Thor., Studi. etc., IV., Vol. 1, p. 77.

1895. „ „ id. The Spiders of Burma, p. 207.

Description of Plate 25 ♀.—*a*, spider, natural size; *b*, profile of spider; *c*, cephalo thorax, underside; *d*, eyes and falces; *f*, epigyne (probably immature).

This curious spider is said by Vinson to live in vertical webs, and he also gives drawings of how it sits on branches which, by its appearance to a growth on the branch, probably serves as a protection. ♂ unknown.

Mr. R. J. Pocock, of the British Museum (at my request) kindly examined the specimen of *Caerostris*, allied to *Paradoxa*, in that collection, and he writes of them as follows:—

"*C. paradoxa* (Dol.), judging from adult ♀ examples from Java and Ceylon in the Museum is perfectly distinct from all the Madagascar species we possess.

"(1) The pale band on the tibiæ covers half the lower surface of the segment, and is yellow. In the others this band covers only one third of the tibia and is quite white.

"(2) The protarsi (metatarsi) are much wider, and the one on the 4th leg is deeply excavated above. In the Madagascar forms the 4th protarsae has no dorsal excavation, and those on the anterior two pairs of legs are slender and have a different curvature.

"(3) The tibia and tarsus of the palp are less expanded, the length of the tibia being nearly twice its width; whereas in the Madagascar species the width is about four-fifths of the length.

"(4) The epigyne is quite different—apart from minor but sufficiently obvious distinctions connected with the shape of the impressions, it may be recognized by the absence of two basally contiguous, apically diverging teeth which project backwards over the impressions in all the Madagascar species known to me.

"In speaking of the Madagascar species I do not include *C. tuberculosa*, (Vins.)"

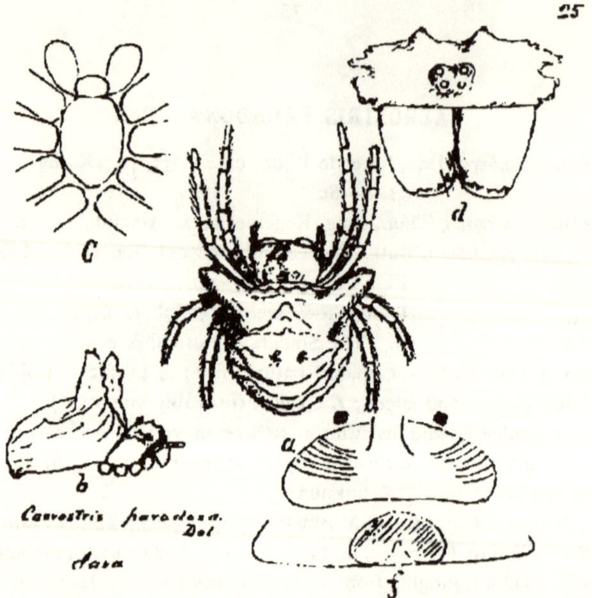
25
C
d
a
b
Caerostris paradoxa.
Dol
Java
f

CAERostris CUSPIDATA. sp n.

Description of Plate 26 ♀.—*a*, figure of spider magnified ; *b*, natural size ; *c*, profile of spider ; *d*, cephalothorax, underside ; *e*, eyes ; *f*, epigyne [hardly mature].

Total length, 4 ; cephalothorax, 1¾ ; breadth, 1½ ; do. in front about 1 ; abdomen, 2¼ ; breadth, 3½ millim. Leg, i.—5, ii.—4¾, iii.—3½, iv.—about 4 millim. Patella + tibia iv.—1 millim.

This little spider, of which I have only one specimen, found by me at Buitenzorg, Java, in 1884, can easily be distinguished by the three curious spines on the top of the cephalothorax. The middle spine is the longest, and the hindmost is double pointed.

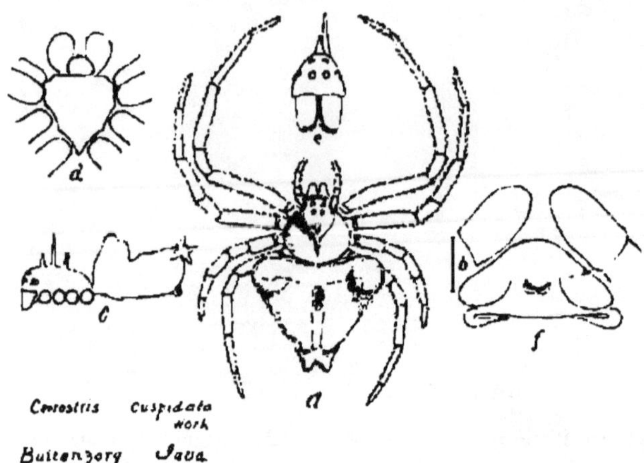
d
e
c
a
b
f
Cerostris cuspidata
Work
Buitenzorg Java

ARGIOPE AEMULA WALCK.

1841. *Epeira aemula*, Walck., H. N.d. Ins. apt., II, p. 118.
1857. „ *(argyopes) striata*, Dol., Bijdr., cet , loc. cit., p. 415.
1859. „ „ „ id, Tweede Bijdr., cet., loc. sit., p. 30, tab. IX., figs. 2, 2a.
1871. *Argiope magnifica*, L. Koch, Die Arachn. Austral., p. 27, plate 11, figs. 6—6b.
1877. „ *aemula*, Thor. Studi, cet., I, Ragni di Selebes, loc. cit., p. 364 (24).
1878. „ „ id. ibid., II, pages 29 and 295.
1881. „ „ id. ibid., III, page 63.
1890. „ „ id. ibid , IV., vol. 1, p. 94.

Description of Plate 27 ♀.—*a*, spider (mag) ; *b*, natural size ; *c*, profile of spider ; *d*, cephalothorax, underside ; *e*, eyes, *f*, epigyne ; *g*, profile of epigyne ; *h*, snare (reduced).

Total length of body, 16 ; cephalothorax, 6¹/₃ ; breadth, 5¼ ; breadth in front, 3²/₇ ; abdomen, 10½ ; breadth, 6½ millim. Legs, i.—30¼, ii.—30, iii.—19, iv.—30 millim. Patella + tibia, iv.—8½ millim. (Thorell).

A. aemula, makes a snare of from 15 to 17 inches in diameter, suspended perpendicularly or at an angle of 45°.

Rays	25—48	
Inner spiral	6—9	turns
Free zone	1	inch
Outer spiral	25—56	turns

Spider sits in centre, head downwards, with its legs stretched out like the letter X. Under the legs white flocculated silk was spun on the web, probably for concealment.

When living this is a very beautiful spider, the colours being most brilliant. It has certainly the power (possessed by several other tropical spiders) of turning quite dark when disturbed. I imagine it is able to do so by raising and depressing the hairy covering of the body.

In many of the species of *Argiope* the white cephalothorax and bands on the abdomen do not show white as long as the specimen is kept in spirits, but appear yellow.

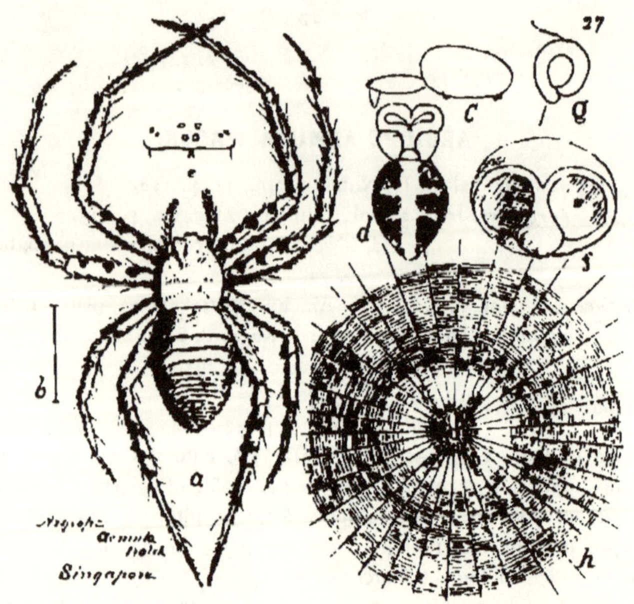
27
c
g
e
d
f
b
a
h
Singapore

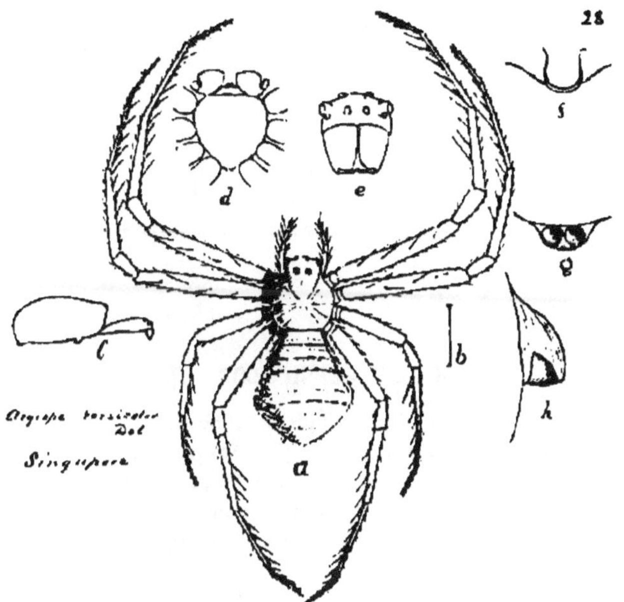
28
d
e
f
g
a
b
c
h
Argiope versicolor Dol
Singapore

ARGIOPE VERSICOLOR. Dol.

1859. *Epeira versicolor*, Dol., Tweede Bijdr., cet., loc. sit., p. 31, tab. IX, fig. 10 (= ♀)
1871. *Argiope succincta*, L. Koch., Die Arachn. Austral., I. p. 35 (= ♀).
1881. „ „ Thor., Studi., cet., III., p. 74, note (= ♀).
1890. „ „ id., ibid., IV., Vol. 1., p. 96.

Description of Plate 28 ♀.—*a*, spider magnified ; *b*, natural size ; *c*, profile ; *d*, cephalothorax underside ; *e*, eyes ; *f*, epigyne ; *g*, do. from behind ; *h*, profile.

Total length, 10½ ; cephalothorax, 4 ; breadth, 4 ; do. in front, 2 ; abdomen, 7 ; breadth, 6 millim. Legs, i.—21, ii.—9¾, iii.—13, iv. ? millim.

I have not observed this spider living, but Doleschall says " It is elegantly coloured and spotted." " A series of bands across the abdomen of different colours—first carmine, then silver, then lemon, then black." " When the ♀ is distended with eggs, these bands are not so evident." " Legs reddish brown colour, but young specimens have dark rings."

ARGIOPE REINWARDTII. Dol.

? 1857. *Epeira (Argyopes) trifasicata*, Dol., Bidjr., etc. p. 416.
1859. „ *reinwardtii*, id., Tweede Bijdr., etc., p. 31, Pl. XV. fig. 5.
1873. *Argiope doleschallii*, Thor. Rem. on Syn., p. 520.
1878. „ „ id. Studi etc. II. pp. 38, 295.
1881. „ „ id. ibid. III. p. 67.
1890. „ *reinwardtii* id. ibid. IV., vol. 1, p. 33.

Description of Plate 29 ♀.—*a*, spider, mag.; *b*, natural size; *c*, profile; *d*, cephalothorax underside; *e*, eyes; *f*, epigyne; *g*, do. profile.

Total length of body, 21; cephalothorax, 8; breadth, 7; breadth in front, 3; abdomen, 13; breadth, 10¾ millim. Legs, i.—46½, ii.—46½, iii.—28½, iv., 45½, millim. long. Metatarsus of 1st pair of legs almost 15, patella+tibia do. 14, patella+tibia IV. pair 12½ millim. long. (Thorell).

A. *reinwardtii*, Dol. makes a perpendicular snare 8 inches in diameter, with a free segment on the upper side. From the centre of the web a line covered with fragments of food projects upwards and backwards like the trap line of other spiders.

Rays	... ...	30
Outer spiral	...	30 turns on upper side.
Do.	...	40 turns on lower side.

29

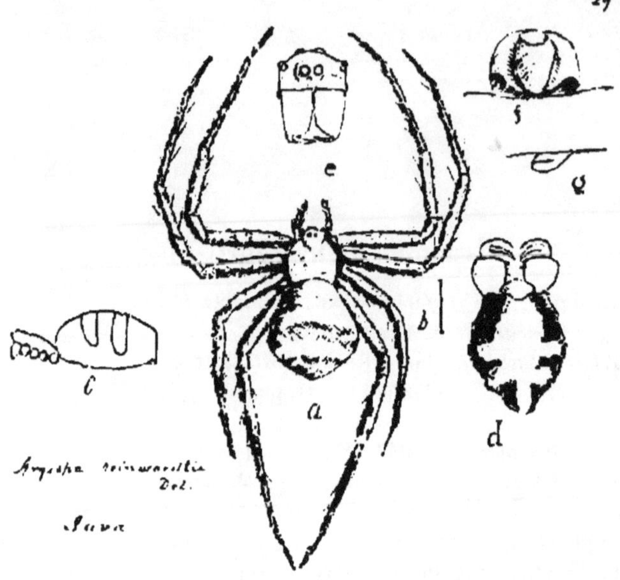
e
f
g
a
b
c
d
Argiope reinwardtii Dol.
Java

GEA FESTIVA. Thor.

1895. *Gea festiva*, Thor. The Spiders of Burma, p. 166.

Description of Plate 30 ♀.—*a*, spider magnified; *b*, natural size; *c*, profile; *d*, cephalothorax underside; *e*, eyes; *f*, epigyne; *g*, do. profile; *h*, snare reduced.

Total length of body, 6; cephalothorax, 2½: breadth, 2¼; in front, 1; abdomen, 4; breadth almost 3¼ millim. Legs, i.—9½, ii.—9¼; iii.—5½; iv.—9 millim. long. Patella + tibia, iv.—3 millim. long. (Thorell).

The snare made by *G. festiva*, Thor. is perpendicular, and from 5 to 6 inches in diameter.

Rays	... ...	48—56
Inner spiral	...	7—11 turns.
Free zone	...	1 inch.
Outer spiral	...	32—38 inches.

In one web the inner spiral spread over the free zone. Another had an orbicular web guarding side of the web on which the spider sat. Spider sits in centre, head downwards, and is easily alarmed, dropping at once to the ground. Spider when living has a bright red ephalothorax. Three specimens found by me at Singapore.

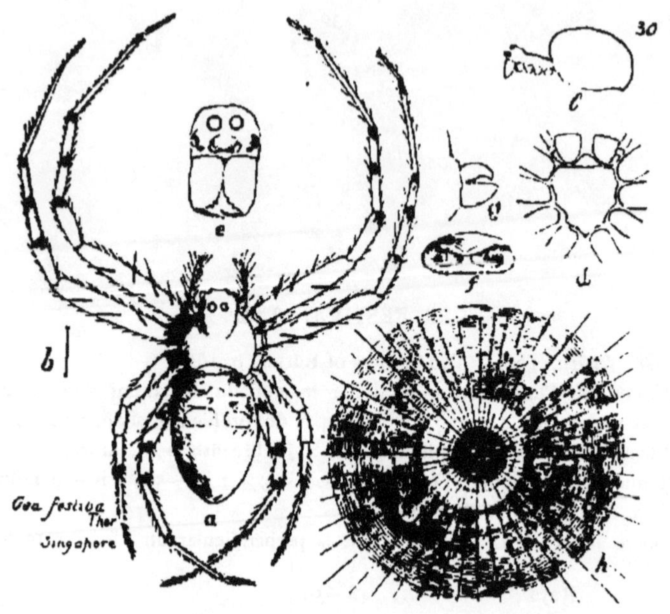
30
c
e
d
f
b
a
h
Gea festiva Thor
Singapore

EUETRIA SALEBROSA. Thor.

1878. *Epeira salebrosa*, Thor., Studi, etc., II pp. 48 and 296.
1881. „ „ id., ibid., III. p. 84.
1890. *Euetria* „ id., ibid., IV., vol. I., p. 115.
1895. „ „ id. The Spiders of Burma, p. 170.

Description of Plate 31 ♀.—*a*, spider magnified; *b*, natural size; *c*, profile; *d*, cephalothorax underside; *e*, eyes; *f*, epigyne; *g*, snare reduced; *h*, coroon.

Total length, 6½; cephalothorax, a little more than 2½; breadth, a little more than 2; in front less than 1; abdomen more than 4; breadth a little more than 2½ millim. Legs, i.—12, ii.—10½, iii.—6½, iv.—11 millim. long. Patella+tibia, iv.—3½ millim. long. (Thorell).

The exceedingly beautiful bell-shaped nest of *E. salebrosa*, Thor. is usually made in grass, and about 3 inches in diameter, with a small round hole at the top. Below the nest there is a protecting screen and above it the cocoons are suspended. Silk of nest is not adhesive. Spider sits underneath at the top, back downwards. When disturbed the white markings on the back of this spider turn quite dark. It is rather a common spider about Singapore.

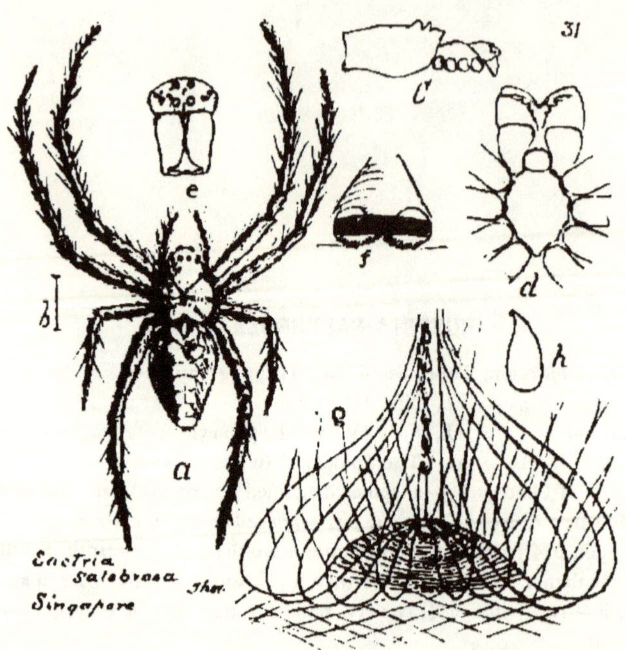
31
C
e
f
d
b
h
a
Eactria
Salebrasa
Thor.
Singapore

EPEIRA CITRICOLA. Forsk.

1775. *Aranea citricola*, Forsk., Descript. Animalium—quae in itinere orient. observ., cet., p. 86.

1776. „ „ id. Icones rer—natural, pl. XXIV. fig. D.

1820. *Epeira opuntiae*, Duf. Ann. gén. sci. phsy. V. p. 355, pl. LXIX., fig. 3.

1841. „ „ Walck. Inst. II. p. 140 Atl. pl. XVIII. fig 2, D et. 2d.

1841. „ *citricola*, id. Ins. Apt. II. p. 143.

184— „ *cacti-opuntiae*, Lucas, in Webb and Berthelot. Hist. Nat. des iles Canaries Anim. Art. Arach., etc., p. 40, pl. VI., fig. 7—7a.

1858. „ *emarginata*, id. in Thomson, Arach. Ent. II., p. 42, pl. 12 ; fig. 5.

1863. „ *flava*, Vinson, Aran. des îsles Réun. Maurice et Madag., pp. 222 and 313, pl. VIII., fig. 3.

? 1872. *Cyrtophora sculptilis*, L. Koch, Die Arachn., Austral, p. 128, pl. IX., fig. 9—9a

1873. *Cyclosa citricola*, Gerst. Von der Decken's Reisen in Ost-Afrika, III. 2, p. 494.

1874. *Cyrtophora opuntiae*, Sim. Les Arachn. de France, I., p. 34, pl. I., fig. 3.

1895. *Epeira citricola*, Thor., The Spiders of Burma, p. 172.

Description of Plate 32 ♀.—*a*, spider magnified ; *b*, natural size ; *c*, profile ; *d*, cephalothorax ; *e*, eyes ; *f*, eight palpus, or eyes ; *g*, epigyne ; *h*, snare reduced.

This spider varies much in size, according to Thorell, from 5¾ to 12 millim.

The snare, 5 inches in diameter, woven by *E. citricola*, has a considerable resemblance to that of *E. beccarii* (Thor.), except that centre of web is but slightly raised, and has much the appearance of a reticulated snare, having innumerable rays spreading out most irregularly. Spiral also irregular. Above the snare is a reticulation, and a flat one below to protect it from beneath.

Snare made horizontally in long grass. Spider sits in centre on lower side, back downwards.

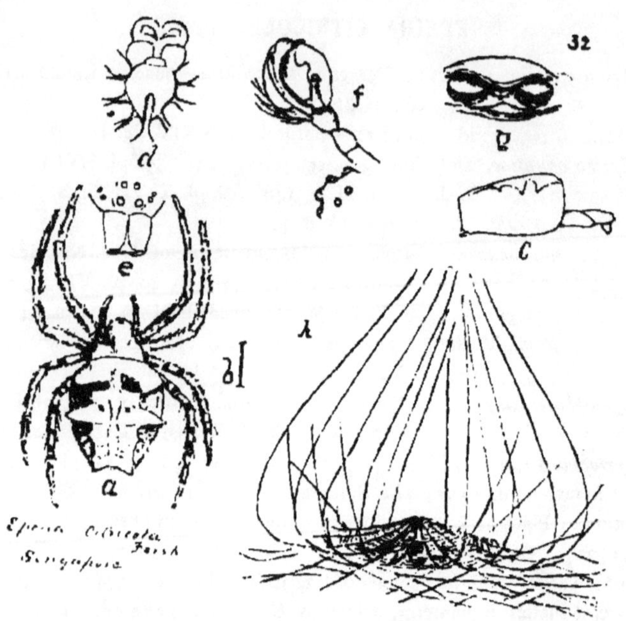
32
d
f
e
c
h
b
a
Epeira Citricola Forsk
Singapore

EPEIRA BIFIDA. Dol.

1859. *Epeira bifida*, Dol., Tweede Bijdr. etc., p. 38, Pl. II., figs. 8—8c (= forma princip.)
1877. „ *macrura*, Thor. Studi, etc., I, p. 64.
1878. „ *bifida*, id., ibid., I., pp. 73 and 297.
1881. „ „ id., ibid., III, p. 124.
1890. „ „ id., ibid., IV. vol. I., p 176.
1895. „ „ id. Spiders of Burma, p. 193.

Description of Plate 33 ♀.—*a*, spider magnified; *b*, natural size; *c*, profile; *d*, cephalothorax underside; *e*, eyes; *f*, epigyne; *g*, do. profile; *h*, left palpus ♂ from below; *i*, do. ♂ from above.

♀. Total length, 11; cephalothorax, $2_{2}/_{3}$; breadth nearly $1_{2}/_{3}$; breadth in front about $5/6$; abdomen, 8½; greatest breadth, 3; least breadth, 1¼ millim.; spinnerett, 2¾ from petiolum and from posterior end 5½ millim. Legs, i.—7½, ii.—6½, iii.—4½, iv.—7 millim. long; patella+tibia, iv.—$2^{2}/_{3}$ millim.

♂. Body, 3¾; cephalothorax about 1¾; breadth more than 1; breadth in front nearly ½; abdomen less than 2; breadth, $1^{1}/_{5}$ millim. Spinnerets, 1¼ from petiolum, and from posterior end about 1 millim. Legs, i.—5, ii.—4¼, iii.—$2^{2}/_{3}$, iv.—$4^{2}/_{3}$ millim. long; patella+tibia less than 1½ millim (Thorell).

Doleschall says that "This spider is common in Amboina, constructing a large regular snare in dark places in the vicinity of buildings, in the middle of which it holds itself, and to which it firmly fastens its flat angular cocoon."

The snares that I have found in Singapore are perpendicular, and from 5 to 7 inches in diameter.

Rays	... ...	40—60.
Inner spiral	...	7—11 turns.
Free zone	...	¼ inch
Outer spiral	...	40—50 turns.

Spider seems never to place itself perpendicularly in the centre of snare, but always more or less sideways, and sometimes has a leaf in the web to which it attaches itself, appearing like a bit of debris. As the webs I found were very small and contained no cocoons, though most of the spiders were mature, I am inclined to think there must be two species.

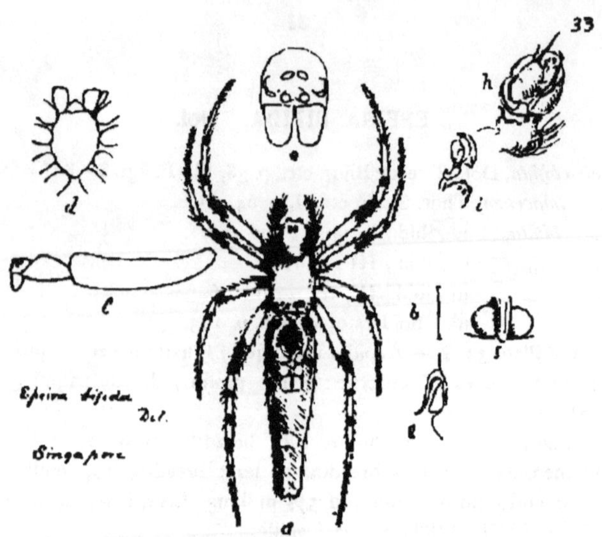
33
Epeira bifida Dol.
Singapore

EPEIRA CONFRAGA. Thor.

1892. *Epeira confraga*, Thor., Novae Species Aranearum, p. 31.

Description of Plate 34 ♀—*a*, spider, mag.; *b*, natural size; *c*, profile; *d*, cephalothorax underside; *e*, eyes; *f*, epigyne; *g*, do. profile; ²/₃*h*, snare reduced.

Total length, 3½; cephalothorax about 1½; breadth, 1; in front at least ½, abdomen, 2½; breadth almost 1½ millim. Legs, i.—3, ii.—2½, iii.—about 2, iv.—almost 3 millim. (Thorell).

This spider's snare is 3 inches in diameter, and perpendicular. Across the web, perpendicularly, is a band of matted grey silk, which is divided at the centre of the web. In this open part the spider sits, head downwards. When at rest it is impossible to distinguish the spider from the silk band, which is constructed, no doubt, for the purpose of concealment.

Rays about	...	...	60.
No inner spiral		...	o ?
Free zone	...	...	½ inch.
Outer spiral about		...	18 turns.

Type specimen in my collection, only one found.

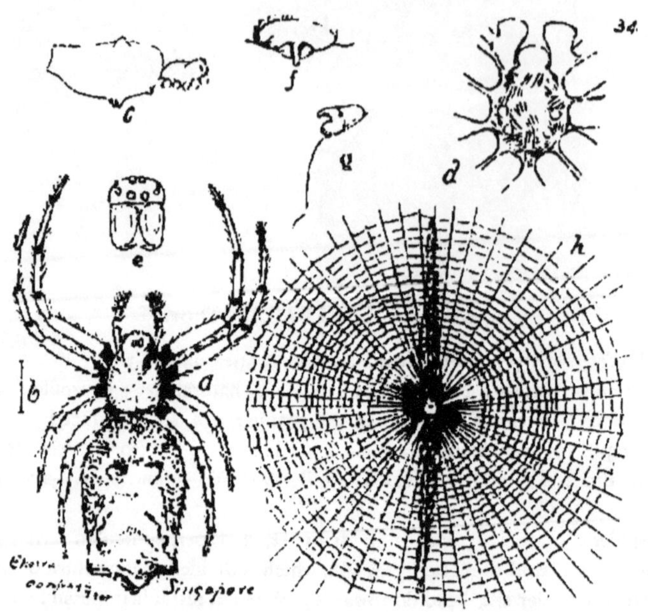
34.
c
f
g
d
e
h
a
b
Singapore

EPEIRA CENTRODES. Thor.

? 1882. *Epeira porcula, var,* Van Hass. Midden Sumatra, etc., p. 22, pl. IV., fig. 6.
1887. „ *centrodes,* Thor. Ragni Birmani, p. 209.
1890. „ „ id. Studi. etc., IV., vol. I., p. 169.
1895. „ „ id. The Spiders of Burma, p. 186.

Description of Plate 35 ♀—*a*, spider mag.; *b*, natural size; *c*, profile; *d*, cephalothorax, underside; *e*, eyes; *f*, epigyne profile; *g*, do. in front

Total length, 7: cephalothorax, $1^5/_6$; breadth, less than 1½; breadth in front, almost 1; abdomen, 6; breadth, 3½ millim. Leg—i. almost 5½, ii.—5⅙, iii. less than 2½; iv.—5 millim. Patella+tibia iv.—1¾ millim. (Thorell).

The snare made by this spider, of which I only obtained one specimen, was perpendicular, 12 inches in diameter.

Rays	... ...	28
Outer spiral	...	57 turns.

Rays do not meet in centre, but have a little open work on which the spider sits head downwards.

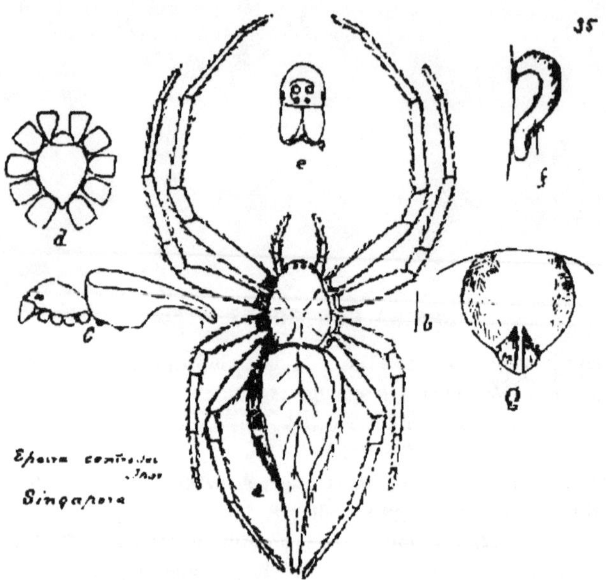
35
d
e
f
c
b
a
Epeira
Singapora

EPEIRA INSULANA. Costa.

1834. *Epeira insulana*, Costa. Cenni. Zool, etc., p. 65.
1841. „ *anseripes*, Walck. H.N. d. Ins. Apt. ii. p. 146.
1842. „ *trituberculata*, Luc. Explor. de l'Algérie, Arachn., p. 248, pl. xv., fig. 4
1877. *Cyrtophora melanura*, Sim. Etudes Arachn. IX., p. 72, pl. iii., fig. 9.
1878. *Epeira anseripes*, Thor., Studi., etc., I., p. 65.
1878. „ „ Thor., Studi., etc., II., p. 405.
1881. „ „ id., Studi., III., p. 124.
1889-90. „ „ id., Studi., IV., p. 175.
1892. „ *insulana*. id. Ann. and Mag. Nat. Hist., p. 232.
1895. „ „ id. The Spiders of Burma, p. 192.
1895. „ „ Pavesi. Annali del Museo, Civico, etc., vol. XXXV., p. 499- Genova.

Description of Plate 36 ♀ —*a*, spider mag.; *b*, natural size; *c*, profile; *d*, cephalo, thorax underside; *e*, eyes; *f*, epigyne; *g*, do. profile; *h*, ♂ left palpus from above; *i*, do. from below.

♀ Total length, 8¾; cephalothorax, 3; breadth, 2⅕; do. in front, 1; abdomen, 6⅕; breadth, 3½ millim. Leg, i.—9½, ii.—9, iii.—6, iv.—9 millim. Patella + tibia iv.—3 millim.

♂ Total length, 5½; cephalothorax, 2½; breadth, 1⅚; do. in front about ⅔; abdomen; 3; breadth, 2 millim. Leg, i.—7½, ii.—7, iii.—5, iv.—almost 7 millim. Patella + tibia, iv.—2¼ millim. (Thorell.)

The snare made by this spider is very variable. In many instances it is a simple circular perpendicular snare of from 5—10 inches in diameter.

Rays	... ...	40—60
Inner spiral	...	7—12 turns.
Free zone		½—1½ inch.
Outer spiral	...	40—60 turns.

The spider sits in the centre on a mass of debris head downwards, though sometimes it places itself horizontally. However, it often makes the snare with a free arc, through which there is a band of grey silk with debris, or with a line of debris across perpendicularly or horizontally, in other cases with a leaf on which it sits.

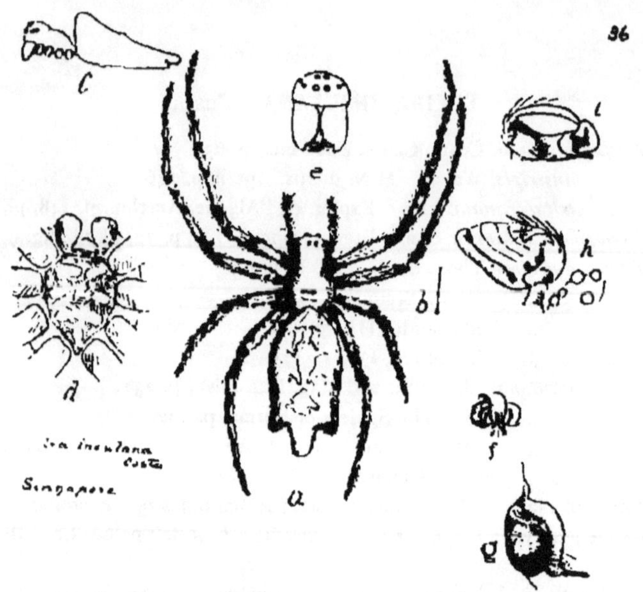
36
c
e
i
h
b
d
a
f
g
insulana
Costa
Singapore

EPEIRA MULMEINENSIS. Thor.

1887. *Epeira mulmeinensis*, Thor., Ragni Birmani, p. 221.

1895. " " id. The Spiders of Burma, p. 192.

Description of Plate 37 ♀—*a*, spider, mag.; *b*, natural size; *c*, profile; *d*, cephalothorax underside; *e*, eyes; *f*, epigyne; *g*, do. profile; *h*, ♂ left palpus, outside; *i*, snare (not enough turns on lower side) reduced; *k*, do. do.

♀ Total length, 4¼; cephalothorax, 2; breadth, almost 1 ⅔; front, almost 5/6; abdomen, 3; breadth, 2½: height, 2⅓ millim. Leg, i.—almost 6, ii.—5, iii.—3½, iv.—almost 5½ millim. Patella + tibia, iv.—1½ millim. (Thorell).

♂ Total length, 3; cephalothorax, 1½; breadth, 1¼; abdomen, 1½; breadth, 1 millim. Leg, i.—3¼; ii.—3; iii.—2; iv.—2½ millim. Patella + tibia, iv.—¾ millim.

This spider, which is very common in Singapore, makes a small perpendicular snare of from 4 to 7 inches in diameter, in which it has a free segment. In this space is a line of grey silk with debris or sometimes cocoons fastened at regular intervals. This line is not of the nature of a trap line as it is fastened to the top suspension line, and the spider lives in centre of web and appears like one of the pieces of debris. In some instances there is debris at lower side of centre. One spider when disturbed shook its web violently.

Rays	...	40—50	
Inner spiral	...	5—7	turns at top.
Free zone	...	¾	inch.
Outer spiral	...	20—30	turns at top.
Do.	...	50—60	do. at bottom.

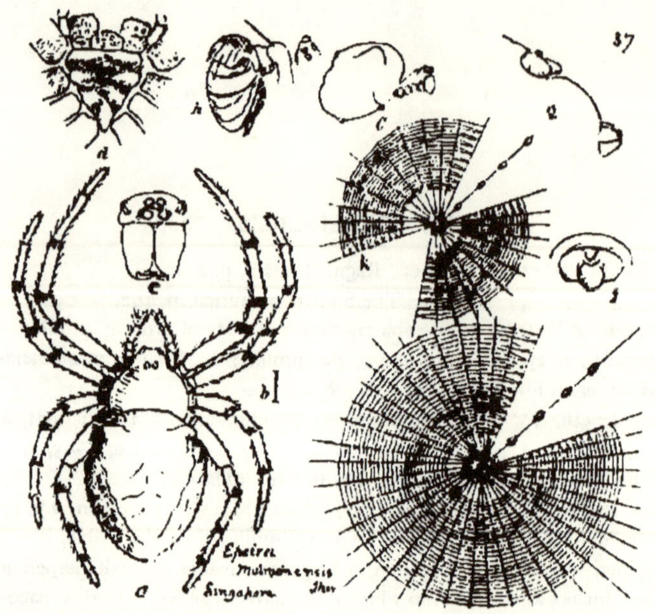
37
Epeira
Singapore
Thor

EPEIRA POSTILENA. Thor.

1878. *Epeira postilena*, Thor., Studi, etc. II., pp. 70, 273, 279.

1881. „ „ id. ibid. III., p. 119.

1895. „ „ id. The Spiders of Burma, p. 178.

Description of Plate 38 ♀—*a*, spider, mag.; *b*, natural size; *c*, profile; *d*, eyes; *e*, maxillae and labium; *f*, epigyne; *g*, do. profile; *h*, snare reduced; *i*, nest.

♀ Total length, 8⅓; cephalothorax, 3½; breadth, 2½; do. in front, 1½; abdomen, 5¾; breadth, 5¼ millim. Leg, i.—13¼; ii.—11¼; iii.—7; iv.—10⁴/₅ millim. Patella + tibia, iv.—3⅔ millim. (Thorell).

♂ Total length, 5; cephalothorax, a little more than 2½; do. in front, almost 1; abdomen, 2½; breadth, a little more than 2 millim. Leg, i—10¾; ii.—almost 9; iii.—5; iv.—7¼ millim. Patella + tibia, iv.—2¼ millim. (Thorell).

This spider, which I have not found in Singapore, and only one specimen in Java, makes a perpendicular snare 3 inches in diameter, with a free segment, with a trap line to nest in a twisted leaf, about 1½ inches long and bound together with silk. Spider seems to sit in nest with its back downwards, holding on by the threads of the roof.

Rays	... ...	15
Inner spiral	...	4 turns.
Free zone	...	¼ inch.

Spider when living is a bright green colour, with black and white markings on the back.

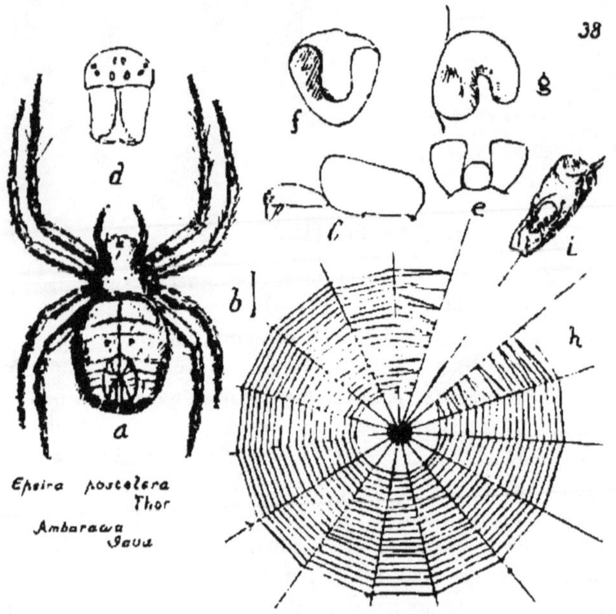
38
d
f
g
C
e
i
b
h
a
Epeira postelera
Thor
Ambarawa
Java

EPEIRA MITIFICA. Thor.

1886. *Epeira mitifica*, Sim. Arachn. de Siam, etc., p. 16.

1887. „ „ Thor. Ragni Birmani, p. 187.

1895. „ „ id. The Spiders of Burma, p. 178.

Description of Plate 39 ♀—*a*, spider, mag.; *b*, natural size; *c*, profile; *d*, maxillae and falces; *e*, eyes; *f*, epigyne, dried; *g*, do., wet; *h*, do., side view, wet; *i*, snare reduced; *k*, nest in leaf.

Total length, 7; cephalothorax, 3; breadth, about 2¼; do. in front, 1¼; abdomen, 5; breadth, 4; height, 3½ millim. Leg, i.—9; ii.—8; iii.—6; iv.—8 millim. Patella + tibia, 2½ millim.

This spider when living is of a beautiful green colour. Snare like *Zilla atrica*, C. L. Koch, but in some snares there are more spirals on lower side than on the upper. Snare, 4 to 5 inches in diameter, and generally extended at an angle of 45°.

Rays ...	...	30—35
Inner spiral	...	6—7 turns.
Free zone	...	¼ inch.
Outer spiral	...	20—30 turns upper side.
Do.	...	23—40 do. lower side.

♂ Much smaller than ♀ but colouring is the same.

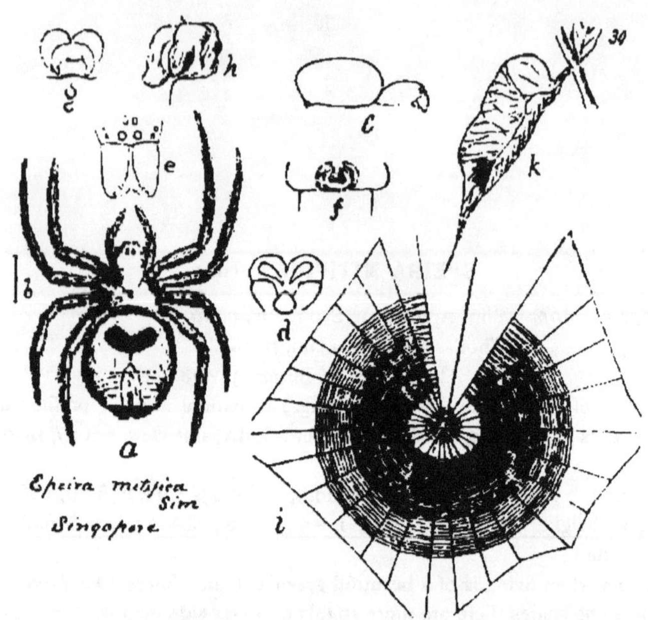
30
c
h
C
e
k
f
b
d
a
i
Epeira mitifica
Sim
Singapore

EPEIRA MICULA. Thor.

1892. *Epeira micula*, Thor. Novae Species, Aranearum, etc., p. 37.

Description of Plate 40 ♀—*a*, spider, mag.; *b*, natural size; *c*, profile; *d*, cephalothorax, underside; *e*, eyes; *f*, epigyne; *g*, do. profile; *h*, snare reduced.

Total length, 2¾; cephalothorax, less than 1; breadth, almost 1; in front, about ½; abdomen, 1½; breadth, about 1¼ millim. Leg, i.—3¼; ii.—almost 3; iii.—1⅚; iv.—a little less than 2½. Patella + tibia, iv.—almost 1 millim. (Thorell).

This spider has great resemblance to *E. mulmeinensis* Thor., but can easily be distinguished from it by its much smaller size, and by its want of abdominal tubercles or black annulations on the legs. Its snare is made on the same plan, perpendicular, from 5 to 8 inches in diameter. There is a free segment, in the centre of which is placed a line of grey silk in which masses of debris or prey is enswathed at regular distances. The spider sits in centre, having the appearance of another mass; in some cases there are other masses at the lower side of centre as in *E. mulmeinensis* Thor.

Rays about	...	60
Inner spiral	...	7 turns.
Free zone	...	½ inch.
Outer spiral	...	12 to 30 turns at top.
Do.	...	30 to 50 do. at bottom.

Type in my collection.

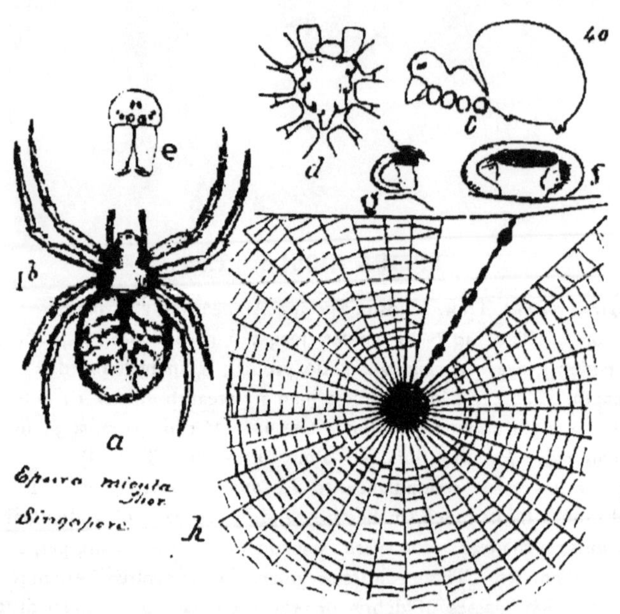
40
e
d
c
f
1b
a
h
Epeira micula Thor.
Singapore

EPEIRA LAGLAIZEI. Sim.

? 1857. *Epeira thomisoides*, Dol. Bijdr., etc, p. 422.
? 1859. „ „ id. Tweede Bijdr, etc., p. Tab. 2, fig. 2.
1877. „ *laglaizei (laglaisei)*, Sim. Études Arachn. IX., p. 65.
1878. „ *thelura*, Thor. Studi., etc., pp. 84, 273, 293.
1881. „ *laglaizei*, id. ibid., III., p. 119.
1890. „ „ id. ibid., IV., vol. I., p. 167.
1895. „ „ id. Spiders of Burma, p. 188.

Description of Plate 41 ♀—*a*, spider, mag.; *b*, natural size; *c*, profile; *d*, cephalothorax, underside; *e*, eyes; *f*, epigyne; *g*, do. profile.

Total length, 10¾; cephalothorax, 3½; breadth, more than 3; do. in front, 1½; abdomen, with tail, 7½; breadth, 6 millim. Leg, i.—14½; ii.—12; iii.—7; iv.—11 millim. Patella+tibia, iv.—3½ millim. (Thorell.)

E. lagiaizei. Sim. makes a perpendicular snare 20 inches in diameter, and sits in centre, head downwards, drawing the rays tightly together with its legs. It is not easily frightened.

Rays	... ...	25	
Inner spiral	...	3	turns.
Free zone	...	2½	inches.
Outer spiral		31	turns at top.
Do.		45	do. at bottom.

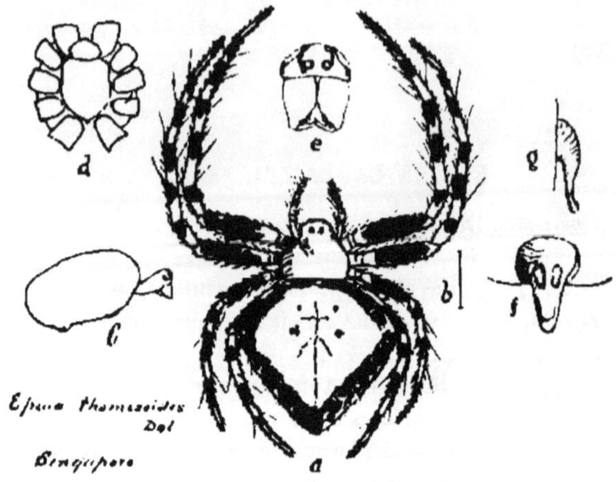
d
e
g
b
f
C
Dol
a

EPEIRA PORCULA. Sim.

1876. *Epeira porcula*, Sim. Études Arachn. IX., p. 78, pl. 3, fig. 7-7a.

1890. " " Thor. Studi. etc., IV., vol. I., p. 172.

Description of Plate 42 ♀—*a*, spider, mag.; *b*, natural size; *c*, profile; *d*, cephalothorax, underside; *e*, eyes; *f*, epigyne; *g*, do. profile.

Total length, 7; cephalothorax, 3½; breadth, 3; do. in front, 1½; abdomen, 5; breadth, 4½; height, 3 millim. Leg, i.—14; ii.—12½; iii.—5; iv.—11 millim. Patella + tibia, iv.—4 millim.

This spider makes a perpendicular web from 18 to 20 inches in diameter.

Rays	17—20	
Innner spiral ...	3—4	turns.
Free zone ...	1—2	inches.
Outer spiral ...	37—42	turns.

Spider sits in centre, head downwards; in one case was sitting under a leaf out of snare.

42

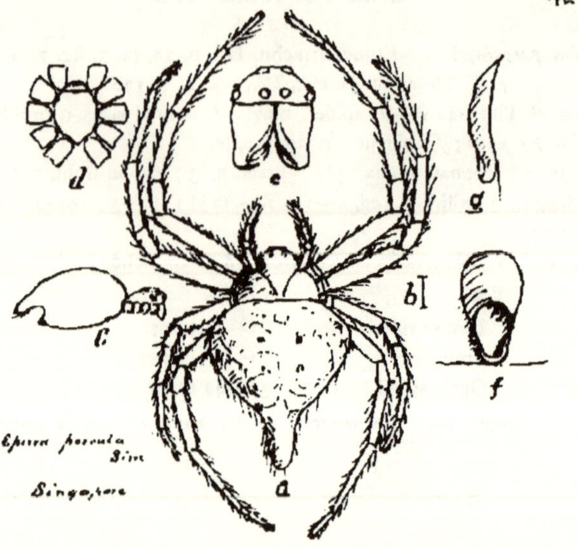
d
e
g
b
c
f
a
Singapore

EPEIRA NOEGEATA. Thor.

1895. *Epeira noegeata*, Thor. The Spiders of Burma, p. 178.

Description of Plate 43 ♀—*a*, spider, mag. ; *b*, natural size ; *c*, profile ; *d*, cephalothorax, underside ; *e*, eyes ; *f*, epigyne in front ; *g*, do. from behind ; *h*, do., side view.

Total length, almost 5¼ ; cephalothorax, almost 2½ ; breadth, 2 ; breadth in front, 1 ; abdomen, almost 3²/₃ ; breadth, almost 4 millim. Leg, i.—about 7¾ ; ii.—7½ ; iii.—3¾ ; iv.—about 5¾ millim. Patella + tibia, iv.—2 millim. (Thorell.)

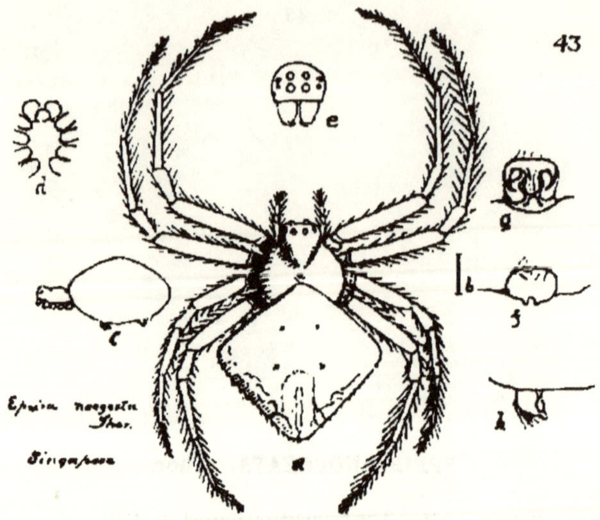
43
e
a
g
b
c
f
h
Epeira
Thor.
Singapora

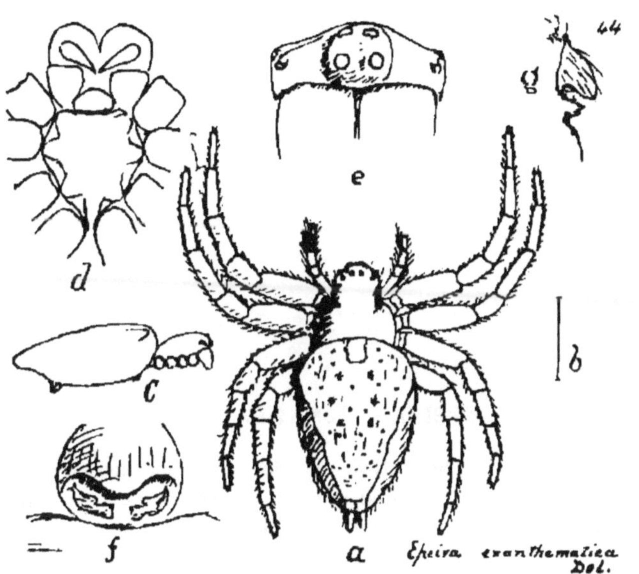
44
g
e
d
b
c
f
a
Epeira exanthematica
Dol.

EPEIRA ECZEMATICA. Thor.

1892. *Epeira eczematica*, Thor. Novae Species Aranearum, etc., p. 23.

Description of Plate 45 ♀—*a*. spider, mag.; *b*, natural size; *c*, profile; *d*, cephalo thorax, underside; *e*, eyes; *f*, epigyne; *g*, do., profile; *h*, cocoon, mag.

Total length, 9½; cephalothorax, 4¼; breadth, 3¼; breadth in front, 2; abdomen, 7; breadth, 4 millim. Leg, i.—11¼; ii.—10½; iii.—7; iv.—9½ millim. Patella +tibia, iv.—3½ millim. (Thorell.)

This spider, with its cocoon, was found in a huge mass of leaves, 12 inches by 6 inches, which it had drawn and matted together with silk. There was no snare. Cocoon yellow, ½ inch in diameter. Type in my collection.

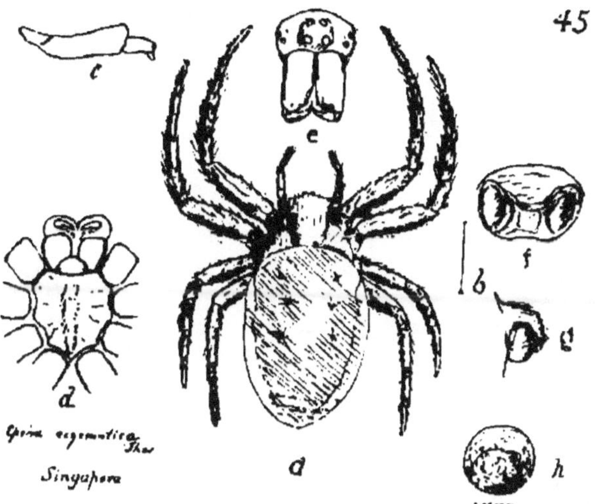
45
c
e
d
a
b
f
g
h
Epeira
Singapore

EPEIRA PERPOLITA. Thor.

1892. *Epeira perpolita*, Thor. Novae Species Aranearum, etc., p. 29.

Description of Plate 46 ♀—*a*, spider, mag.; *b*, natural size; *c*, profile; *d*, cephalothorax, underside; *e*, eyes; *f*, epigyne.

Total length, 5; cephalothorax, 2; breadth, about 1½; breadth in front, about ¾; abdomen, less than 4; breadth, 3½ millim. Leg, i.—6¾; ii.—5¾; iii.—3½; iv.—5 millim. Patella+tibia, iv.—almost 2 millim. (Thorell.)

Type in my collection.

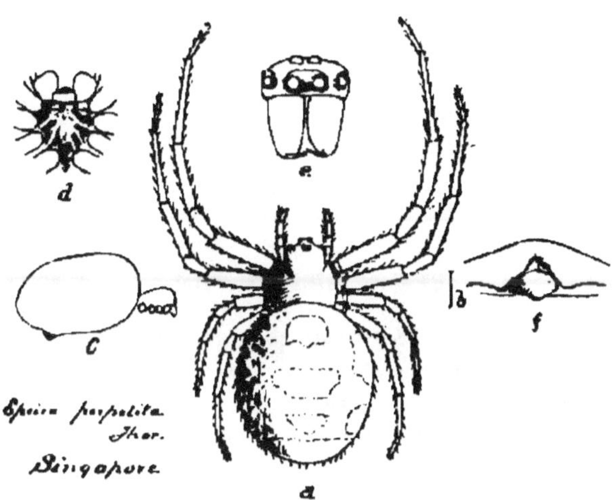
d
e
C
f
a
Epeira perpolita
Thor.
Singapore

EPEIRA HISPIDA. Dol.

1859. *Epeira hispida*, Dol. Tweede Bijdr., p. 33, pl. 2, fig 5 (= ♀).
1877. " *decens*, Thor. Studi. I., etc., p. 379 (39) (= ♂).
1878. " *rumpfii*, id. ibid. II., p.296 (= ♂).
1884. " *rufofemorata*, Sim. Arachn. rec. in Birmanie par Comotto, p. 348 (24). (= ♀).
1887. " *hispida*, Thor. Ragni Birmani, p. 179.
1895. " " id. The Spiders of Burma, p. 177.

Description of Plate 46 ♀—*a*, spider, mag.; *b*, natural size; *c*, profile; *d*, cephalothorax, underside; *e*, eyes; *f*, epigyne; *g*, do., profile.

♀ *Epeira hispida*, varies greatly in size, but in most specimens the total length is 13; cephalothorax, almost 6; breadth, 5; abdomen, 9; breadth, 7¾ millim. Leg, i.—21; ii.—20; iii.—13; iv.—19 millim. Patella + tibia, iv.—almost 7 millim. (Thorell).

♂ Total length, 9; cephalothorax, 5; breadth, 4; breadth in front, 1½; abdomen, 5½; breadth, scarcely 4 millim. Leg, i.—17½; ii.—15; iii.—10; iv.—14 millim. Patella + tibia, iv.—5 millim. (Thorell).

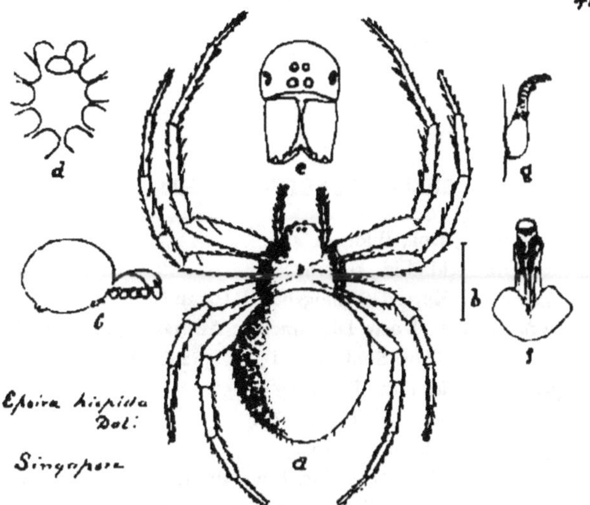
d
e
g
b
c
f
a
Epeira hispida
Dol.
Singapore

EPEIRA PUNCTIGERA Dol.

1857. *Epeira punctigera*, Dol., Bijdr., p. 420.
1857. „ *manipa*, id. ibid., p. 419.
1863. „ *triangula*, Keyserl. Sitzungsber. d. Isis zu Dresden, p. 98, pl. V, figg. 12-14.
1871. „ *indagatrix*, L. Koch. Die Arachan. Austral. (1) p. 66, pl. V., figg. 8-9a.
1877. „ *vatia*, Thor. Studi. etc., I, pp. 382 & 384.
1878. „ *punctigera*, id. ibid., II., p. 59 & 256.
1881. „ „ id. ibid., III. p. 104.
1881. „ *ephippiata*, id. ibid., p. 101.
? 1886. „ *pavicei*, Sim. Actes Soc. Linn. Bordeaux, XL., p. 150.
1887. „ *punctigera*, Thor. Ragni. Birmani., p. 181.
1890. „ „ id. ibid. Studi., IV., vol. I., p. 147.
1895. „ „ id. The Spiders of Burma, p. 177.

Description of Plate 48 ♀—*a*, spider, mag. ; *b*, natural size ; *c*, profile ; *d*, cephalothorax, underside; *e*, eyes ; *f*, epigyne ; *g*, do., profile ; *h*, ♂ right palpus, side view ; *i*, do. from below.

♀ Length of body, 10¾ ; cephalothorax, 4¾ ; breadth, 3¾ ; breadth in front, 2¹/₅ ; abdomen, 7¹/₅ ; breadth, 7 millim. Leg, i.—17¼ ; ii.—14½ ; iii.—9½ ; iv.—14½ millim. Patella + tibia, iv.—almost 5 millim. Falces, 2 millim. long. (Thorell).

♂ Length of body, 9 ; cephalothorax, 5 ; breadth, 4 ; breadth in front, 1½ ; abdomen, 5 ; breadth, 3⁴/₅ millim. Leg, i.—19½ ; ii.—17½ ; iii.—10 ; iv.—16 millim Patella + tibia, iv.—5¼ millim. (Thorell).

Found in dauber wasp's nest.

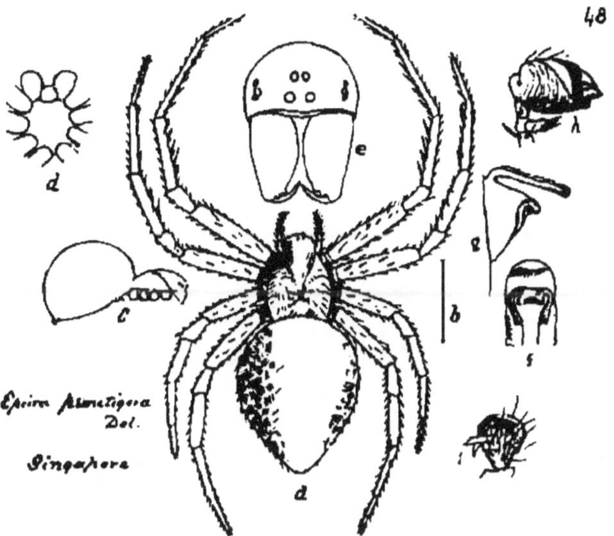
48
b
e
d
c
b
g
f
h
i
d
Epeira
Dol.
Singapore

EPEIRA ANASPASTA. Thor.

1892. *Epeira anaspasta*, Thor. Novae Species Aranearum, etc., p. 26.

Description of Plate 49 ♀—*a*. spider, mag.; *b*, natural size; *c*, profile; *d*, cephalothorax, underside; *e*, eyes; *f*, epigyne; *g*, do., profile.

Total length, 11½; cephalothorax, 5; breadth, hardly 4; breadth in front, hardly 2; abdomen, 6½; breadth, about 5¾ millim. Leg, i.—19¼; ii.—17¾; iii.—10½; iv.—15½ millim. Patella+tibia, iv.—5½ millim. (Thorell.)

Found in a dauber wasp's nest.

Type in my collection.

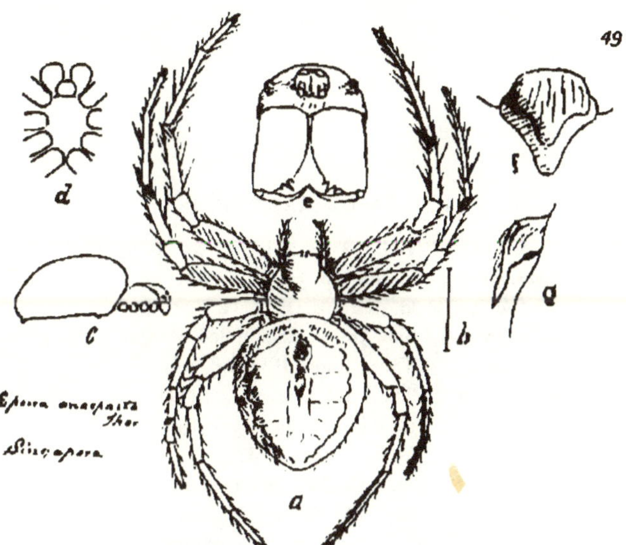
49
d
e
f
c
b
g
a
Epeira anaspasta
Thor
Singapora

EPEIRA DE HAANII. Dol.

	1859.	*Epeira de haanii,*	Dol. Tweede Bijdr., etc., p. 33, pl. II., fig. 7.
?	1859.	„ *bogoriensis,*	id. ibid., p. 35, pl. XI., fig. 7.
	1859.	„ *spectabilis,*	id. ibid , p. 34, pl. II., figg. 9-9b.
	1887.	„ *submucronata,*	Sim. Journ. Asiatic Soc. Bengal, LVI., p. 106.
	1887.	„ *de haanii,*	Thor. Ragni Birmani., p. 178.
	1877.	„ *kandarensis,*	Thor. Studi., I , p. 372 (32).
	1878.	„ *de haanii,*	id. ibid., II., pp. 25 & 296.
	1881.	„ „	id. ibid., III., p. 88.
	1890.	„ „	id. ibid., IV., vol. 1, p. 125.
	1895.	„ „	id. The Spiders of Burma, p. 170.

Description of Plate 50 ♀—*a*, spider, mag.; *b*, natural size; *c*, profile; *d*, cephalothorax, underside; *e*, eyes; *f*, epigyne; *g*, do., profile; *h*, do., from behind.

Total length, 25¾ : cephalothorax, 9⅓; breadth, 7¾; do. in front, 4⅓; abdomen, 16¾; breadth, 11 millim. Leg, i.—31; ii.—30; iii.—19½; iv.—32½ millim. Patella+tibia, iv.—11½ millim Falce, 4 millim. long. (Thorell.)

Doleschall says this spider is found in great numbers in the clove gardens at Amboina during the rainy season, but I did not find it common at Singapore.

It makes a large perpendicular snare, from 18-24 inches in diameter, and has one of the rays carried out to its nest under a leaf.

In the centre of the web is a large hole, round which the inner spiral irregularly winds two or three times. Free zone about 1 inch. Outer spiral from 25 to 30 turns. Free zone about 1½ inches. Rays about 20 in number, and very irregularly spaced.

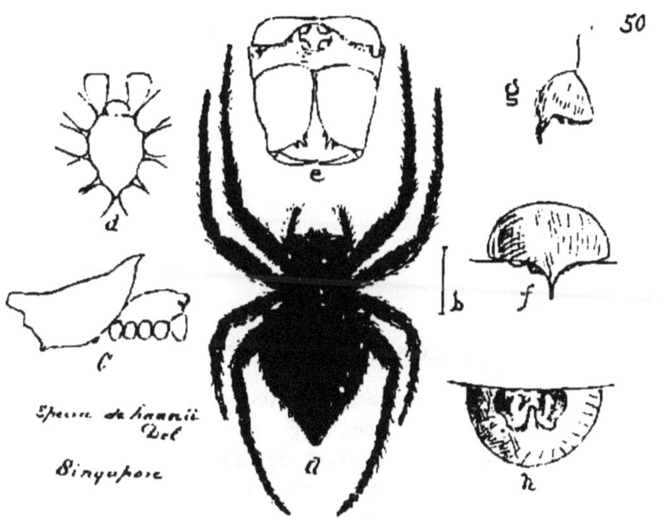
50
g
e
d
f
b
c
a
h
Epeira dehaanii
Dol
Singapore

MILONIA OBTUSA. Thor.

1892. *Milonia obtusa*, Thor. Novae Species Aranearum, p. 40.

Description of Plate 51 ♀—*a*, spider, mag ; *b*, natural size ; *c*, profile ; *d*, cephalo thorax, underside ; *e*, eyes ; *f*, epigyne ; *g*, snare, reduced.

Total length, 5 ; cephalothorax, 2¼ ; breadth, 1⅔ ; breadth in front, 1 ; abdomen, 3 ; breadth, 2 millim. Leg, i.—not quite 5 ; ii—about 4½ ; iii.—a little less than 3 ; iv.—4²/₃ millim. Patella + tibia, iv.—a little less than 1½ millim. (Thorell).

This spider, of which I only found one specimen, constructs a perpendicular snare of about 12 rays, with a trap line to its nest under a leaf. Central spiral, incomplete, starts from an interlaced part, from which trap line leads to nest.

Outer spiral, consisting of 36 threads, is incomplete, and is on the opposite side of the snare from the trap line. Outer spiral adhesive.

Type in my collection.

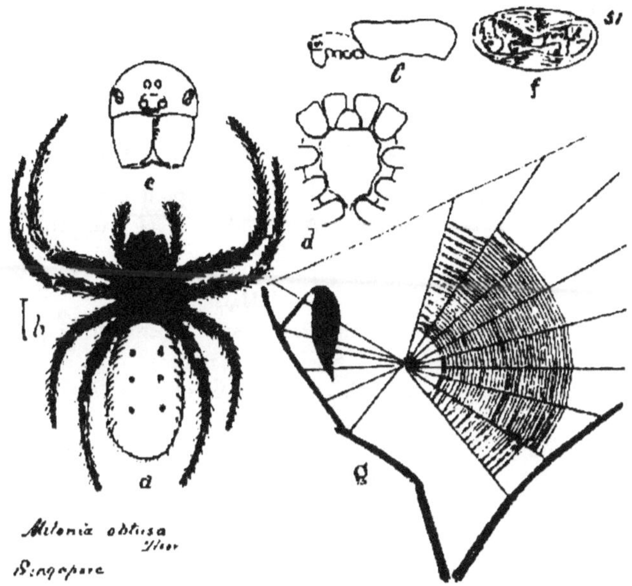
51
c
e
f
d
b
a
g
Milonia obtusa
Singapore

ARGYROEPEIRA CELEBESIANA WALCK.

1841.	*Tetragnatha celebesiana*,	Walck.	H.N. d. Ins. Apt. ii. p. 222.
1864.	" *decorata*,	Blackw.	Ann. and Mag. of Nat. Hist., 3 Ser., XIV., p. 44.
1869.	" "	Cambr.	The Linn. Soc. Journ., Zool. X., p. 389, pl. XIII., figg. 61-68.
1872.	*Meta* "	L. Koch.	Die Arachn. Austral. p. 141, pl. XI., fig. 5.
1877.	" *celebesiana*,	Thor.	Studi I., p. 422 (82).
1878.	" "	id.	ibid. II., pp. 91 & 297.
1881.	" "	id.	ibid. III., p. 126.
1887.	*Argyroepeira* "	id.	Ragni Birmani, p. 138.
1890.	" "	id.	Studi., IV., vol. I., p. 198.
1895.	" "	id.	The Spiders of Burma, p. 155.

Description of Plate 52 ♀—*a*, figure, mag.; *b*, natural size; *c*, profile; *d*, eyes; *e*, maxillae and falces; *f*, epigyne.

Total length, 9½; cephalothorax, 2½; breadth, 2½; breadth in front, 1¾; abdomen, 7; breadth, 3½; height, 4 millim. Leg, i.—about 18; ii.—14; iii.—about 7; iv.— 11 millim. Patella+tibia, iv.—3½ millim.

This spider makes, in low shrub, a rather irregular oval-shaped snare, 9 inches in diameter, and suspended at an angle of 45° to the plane of the earth.

Rays, ...	18.
Inner spiral, ...	6 turns.
Free zone, ...	2 inches,
Outer spiral, ...	20 turns.

Spider sits in the centre. At one side near the centre is a large hole, seemingly made after the snare was completed, and may have been for the spider to get from side to side of the web.

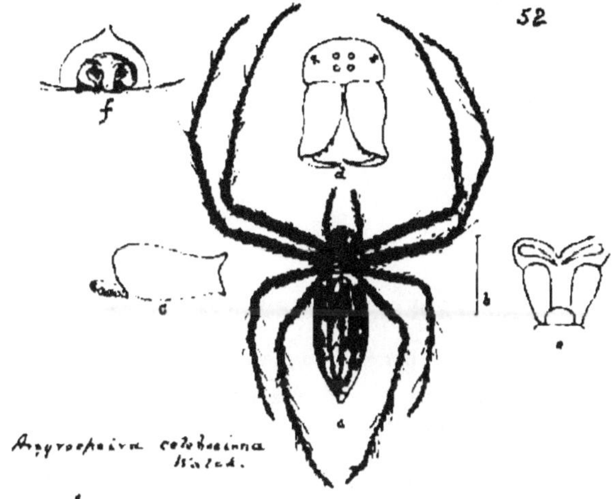
52
f
d
c
b
e
a
Argyroepeira celebesiana
Walck.
Java

ARGYROEPEIRA FIBULATA. Thor.

1892. *Argyroepeira fibulata*, Thor. Nova Species Aranearum, p. 16.

Description of Plate 53 ♀—*a*, spider, mag. ; *b*, natural size ; *c*, profile ; *d*, cephalothorax, underside ; *e*, eyes ; *f*, epigyne.

Total length, 2¾ ; cephalothorax, nearly 1 ; breadth, about ¾ ; breadth in front, almost ½ ; abdomen, 2½ ; breadth, almost 2 millim. Leg, i.—6 ; ii.—less than 5 ; iii.—about 2½ ; iv.—almost 4 millim. Patella+tibia, iv.—about 1 millim. (Thorell.)

This handsome spider, of which I found two specimens, constructs a circular horizontal snare about 4 inches in diameter, and sits in centre on lower side.

Rays	40.
Inner spiral,	7 turns.
Free zone, ...	½ inch.
Outer spiral, ...	25 in No. 1 web.
do.	36 in No. 2 do.

Spider can easily be known by the fibula-like white marking on the lower side of the abdomen in front of the spinnarets.

Type in my collection.

53

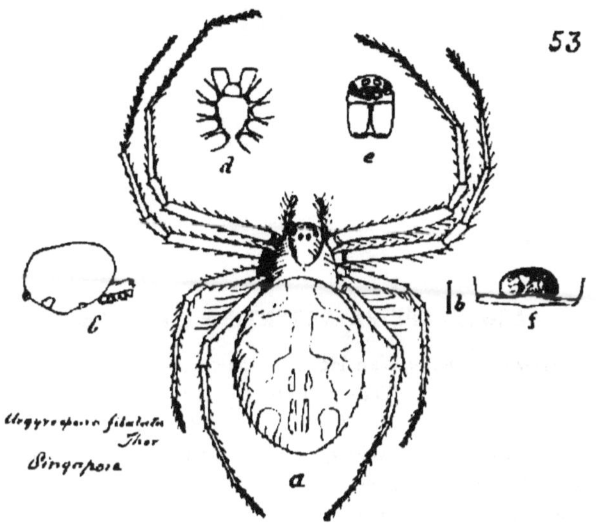

ARGYROEPEIRA ARGENTINA. Van Hass.

1871. *Argyroepeira argentina*, Van Hass. Araneae exoticae, etc., p. 218 (2).
1890. „ „ Thor. Studi., etc., IV., vol. I, p. 199.
1895. „ „ id. The Spiders of Burma, p. 152.

Description of Plate 54 ♀—*a*, spider, mag.; *b*, natural size; *c*, profile; *d*, cephalothorax, underside; *e*, eyes; *f*, epigyne.

Total length, 4½; cephalothorax, almost 1⅚; breadth, almost 1½; breadth in front ¾; abdomen, 3¼; breadth, a little more than 2 millim. Leg, i.—8⅓; ii.—scarcely 7; iii.—3½; iv.—6 millim. Patella+tibia, iv.—almost 1⅚ millim. (Thorell).

Argentina Van Hass always suspends its snare, which is 9-12 inches in diameter, at an angle of 45°, and sits in centre on the lower side.

Rays meeting in ⅛ inch circle, 35-40.
Inner spiral, ... 2-4 turns.
Free zone, ... 1-1¼ inches.
Outer spiral, ... 55-60 turns.

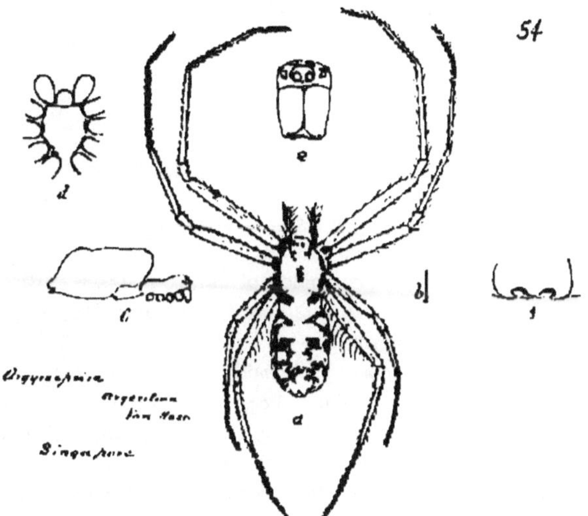
54
d
e
c
b
f
a
Singapore

ARGYROEPEIRA VENTRALIS. Thor.

1877. *Meta ventralis*, Thor. Studi, etc., i., p. 423 (83).
1887. *Argyroepeira* „ id. Ragni Birmani. etc., p. 138.
1895. „ „ id. The Spiders of Burma, p. 155.

Description of Plate 55 ♀—*a*, spider, mag.; *b*, natural size; *c*, profile; *d*, eyes; *e*, cephalothorax, underside; *f*, epigyne.

Total length, $7^1/_5$; cephalothorax, 3; breadth, 2; do. in front, $1^1/_6$; abdomen, 5; breadth, $2^2/_3$; height, more than 3 millim. Leg, i.—22½; ii.—$18^1/_3$; iii.—7: iv.—12 millim. Patella+tibia, iv.—$3^1/_3$ millim. Falce, $1^2/_3$ millim. long. (Thorell.)

This spider, which is very common at Singapore, makes an extremely variable snare, The snares are suspended at any angle from horizontal to perpendicular, and in some cases have a free segment, and in others the circle is complete.

Where the spider makes a free segment it has generally a trap line, leading to its nest close to the snare, which it holds firmly, the legs being stretched out in a line; sometimes the head being towards the snare, but often not.

When the spider constructs a complete orbicular snare horizontally it often remains on the under side in the centre, and in one case when it ran out to capture a fly it dropped down and then climbed up the line, which it spun as it went out, to the centre again.

Snare from 5—15 inches in diameter.
Rays ... 20—30, meeting in ¼ inch circle.
Inner spiral, 2—4 turns.
Free zone, 1—1¼ inch.
Outer spiral, 20—33 turns, adhesive.

Usually the free segment is in the upper quarter, but in one case it was in the lower quarter. The outer spiral has more threads on the side opposite the free segment than at the free segment.

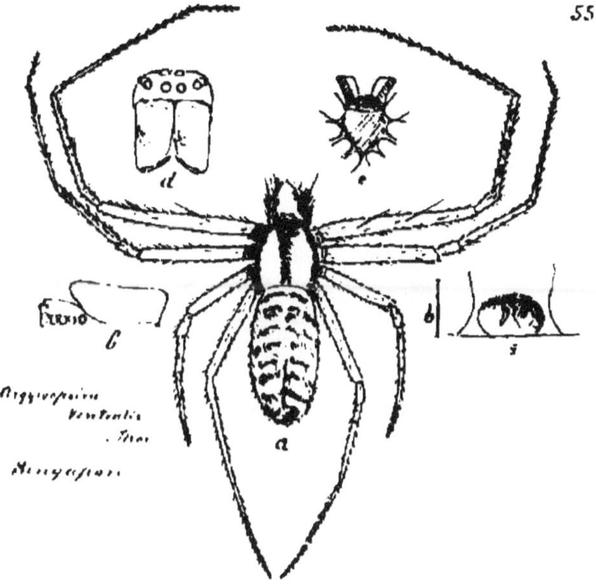
d
e
a
b
C

ARGYROEPEIRA GEMMEA. Van Hass.

1882. *Meta gemmea*, Van Hass, Midden-Sumatra, etc. Araneae, p. 26, pl. ii., fig. 4.

1890. *Argyroepeira* „ Thor., Studi., etc., IV., vol. I., p. 206.

1895. „ „ id. The Spiders of Burma, p. 152.

Description of Plate 56 ♀ —*a*, spider, mag.; *b*, natural size; *c*, profile; *d*, cephalothorax, underside; *e*, eyes; *f*, epigyne.

Total length, 5½; cephalothorax, 2; breadth, 1½; breadth in front, almost 1; abdomen, 4; breadth, 2 1/5 millim. Leg, i.—11; ii.—9; iii.—4 1/3; iv.—a little more than 7 millim. Patella+tibia, iv.—2 millim. (Thorell).

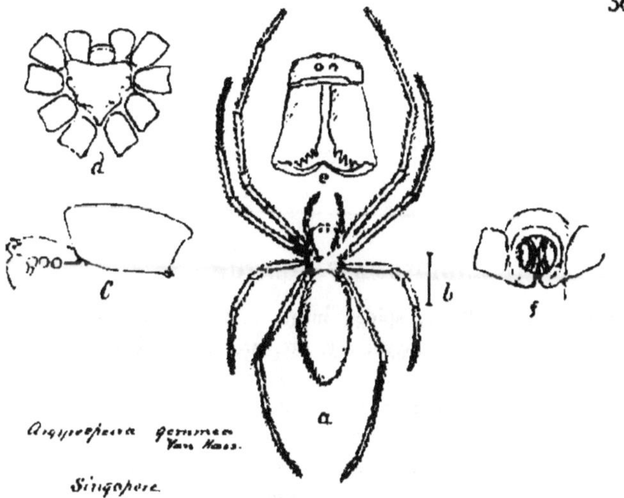
d
e
C
b
f
a
gemmea Van Hass.
Singapore

THERIDIOSOMA FASCIATUM. sp n.

Description of Plate 57 ♀—*a*, spider, mag.; *b*, natural size; *c*, profile; *d*, cephalothorax, underside; *e*, eyes; *f*, epigyne; *g*, do., profile; *h*, ♂ left palpus from below; *i*., do. side view; *k*, snare.

♀ Total length, 2; cephalothorax, 1; breadth, 1; do. in front, ½; abdomen, 1½; breadth, 1½; height, 1½ millim. Leg, i.—2; ii—1½; iii.—1; iv.—1¼ millim. Patella + tibia, iv.—½ millim. As the spider is so small, measurements are only approximate. ♀ Though much resembling ♂, is considerably smaller than it. *T. fasciatum* is closely allied to *T. argenteolum* Cambr., the type of the genus, and *T. radiosa*, M'Cook, which Prof. Thorell considers to be quite identical with *T. gemmosum* L. Koch.

This spider constructs an orb web, though rather an imperfect one, and then draws out the centre of the web and holds it out taut with a trap line until an insect becomes entangled, when it suddenly lets go, thereby more securely holding its prey, as described by Dr. M'Cook in his exceedingly interesting work on "American Spiders and their Spinning Work, vol. 1, page 195."

It is not uncommon in the Botanic Gardens, Singapore.

Type in my collection.

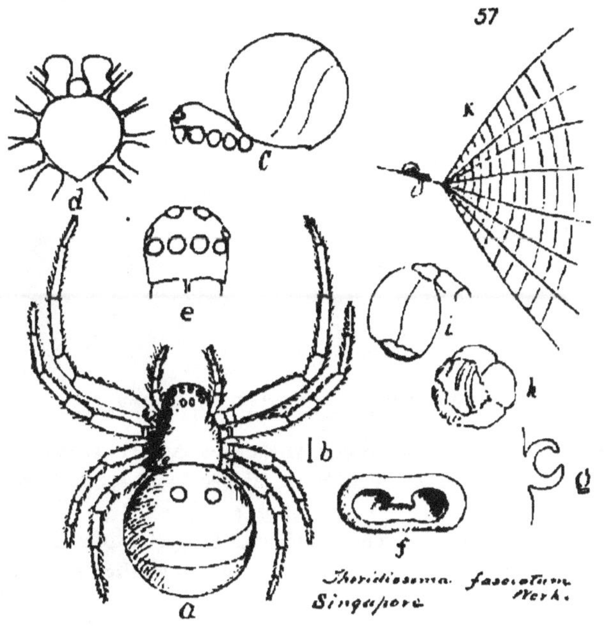
57
a
b
c
d
e
f
g
h
i
k
Theridiosoma fasciatum
Work.
Singapore

PHORONCIDIA LYGEANA. Walck.

1841. *Plectana lygeana*, Walck. H.N.d. Ins. Apt. II., p. 197.

1882. *Phoroncidia acrosomoides*, Van. Hass., Midden Sumatra, etc., Araneae, p. 30, pl. 1, fig. 7, pl. 4, figg. 2 and 3.

1890. „ *lygeana*, Thor. Studi, etc., IV., vol. I, p. 243.

Description of Plate 58 ♀—*a*, spider, mag.; *b*, natural size; *c*, profile; *d*, cephalothorax, underside; *e*, eyes; *f*, epigyne.

Total length, 5: cephalothorax, almost 2; breadth, almost 1¾; breadth of clypeus, about ²/₃; abdomen, 3½: breadth, slightly more than 2½ millim. Leg, i.—6¾; ii.—almost 4; iii.—3¾; iv.—about 6½ millim. Patella + tibia, iv.—2¼ millim. (Thorell.)

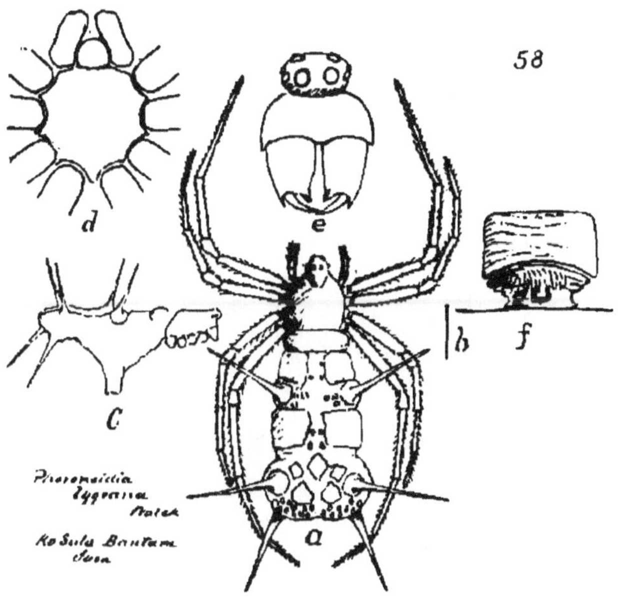
58
d
e
f
b
C
a

MOLIONE TRIACANTHA. Thor.

1892. *Molione triacantha*, Thor., Novae Species, Aranearum, p. 8.

Description of Plate 59 ♀—*a*. spider, mag.; *b*, natural size; *c*, profile; *d*, cephalothorax, underside; *e*, eyes; *f*, epigyne; *g*, spine.

Total length, a little more than 2; length and breadth, about ¾; abdomen, almost 1½; breadth, 1¼ millim. Leg, i.—4 (tibia almost 1); ii —3; iii.—about 2¼; iv.—almost 3¾ millim. Patella+tibia, iv.—almost 1 millim. (Thorell.)

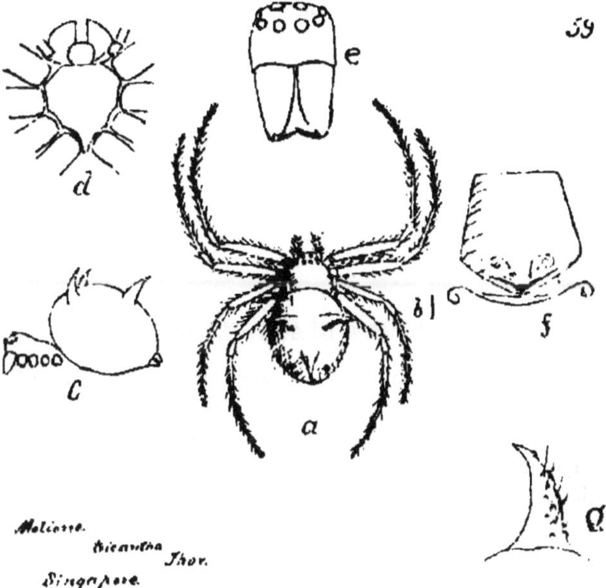

Meliorne.
bicantha Thor.
Singapore.

COLEOSOMA BLANDUM. Camb.

1882. *Coleosoma blandum*, Cambr., Proc. Zool. Soc., London, p. 428, pl. xxix., figg. 3a—3f.

1895. „ „ Thor., The Spiders of Burma, p. 115.

Description of Plate 60 ♂—*a*, spider, mag.; *b*, natural size; *c*, profile; *d*, cephalothorax, underside; *e*, right palpus; *f*, do. from outside; *g*, pitcher plant, *Nepenthes ampullacea*.

Total length, almost 2½; cephalothorax, about 5/6; breadth, almost 2/3; abdomen, 1½; breadth, almost 2/3 millim. Leg, i.—4 2/3; ii.—a little more than 3; iii.—2½; iv.—4 millim. Patella+tibia, iv.—nearly 1½ millim. (Thorell.)

The snare of this little spider was found spun across the mouth of one of the small pitcher plants locally called "Monkey Cups," which are to be seen growing wild in considerable numbers in the Botanic Gardens, Singapore. Though I found several webs I was only able to obtain one specimen of the spider.

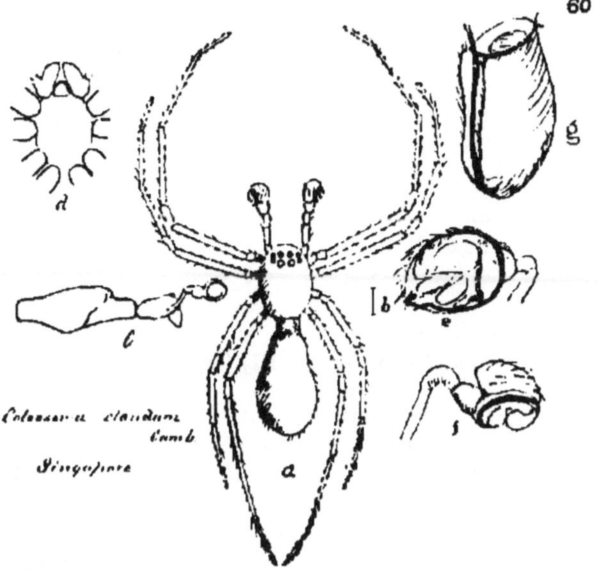
60
g
d
b
e
c
f
a
Coleosoma blandum
Camb
Singapore

ARIAMNES FLAGELLUM. Dol.

1857. *Ariadne flagellum*, Dol. Bijdr, etc., p. 411, fig 1.

1895. *Ariamnes* „ Thor. The Spiders of Burma, p. 74.

Description of Plate 61 ♀—*a*, spider, mag.; *b*, natural size; *c*, cephalothorax underside; *d*, eyes.

Total length, 24; cephalothorax, 2; breadth, 1; abdomen, 22; breadth, 1 millim. Leg, i.—?; ii.—8; iii—5; iv.—18 millim. Patella+tibia, iv—4½ millim.

Doleschall says that "*A. flagellum* Dol. when alive has the cephalothorax and abdomen grass green, slightly yellow in the middle, and upon both sides are numerous small gold sparkling speckles." I have not seen this species living, but an allied species, probably *A. attenuata* Camb. found by me in Brazil is also when alive of a bright grass green, and when it is seen suspended by a thread from its spinnerets is remarkably like a fragment of grass.

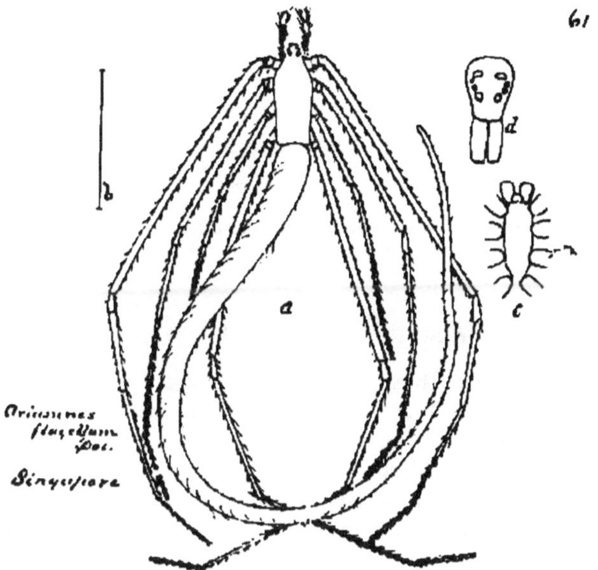
61
b
d
c
a
Ariamnes
flagellum
Poc.
Singapore

ARGYRODES XIPHIAS. Thor.

1887. *Argyrodes xiphias*, Thor. Ragni Birmani, p. 95.

1895. „ „ Thor. The Spiders of Burma, p 118.

Description of Plate 62 ♀—*a*, spider, mag; *b*, natural size; *c*, profile; *e*, cephalothorax, underside; *f*, eyes and falces; *g*, do. profile; *h*, epigyne; *i*., do. profile; *a*, ♂ profile; *k*, ♂ right palpus, from outside; *l*, do. in front.

♀ Total length, scarcely 3; cephalothorax, 1; breadth, about ²/₃; abdomen from base to posterior apex, 2; height, almost 2; breadth, 1¼ millim. (Thorell). Leg, i.—about 4; ii.—about 2; iii.—about 1¾; iv.—about 1½ millim.

♂ Total length without horn, 2¾; cephalothorax, less than 1; breadth, about ¾; abdomen, 1¾; breadth and height about ²/₃ millim. Leg, i.—(about 6); ii.—3½; iii.—about 1⅓; iv.—about 3 millim. Patella+tibia, iv.—about ⁵/₆ millim. (Thorell).

A. xiphias Thor. lives in the webs of *Nephila maculata*, Fabr.

A. nasuta Camb. closely resembles this spider, but has the projection in front of the clypeus somewhat spoonshaped at its extremity, while in this species there is no dilation.

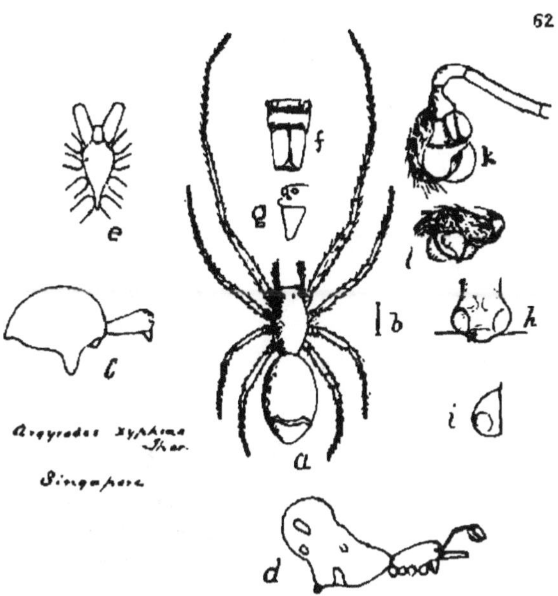
e
f
g
k
l
b
h
c
i
a
Argyrodes xyphias
Thor.
Singapore
d

ARGYRODES FASCIATUS. Thor.

1892. *Argyrodes fasciatus*, Thor. Novae Species Aranearum, etc., p. 4.

Description of Plate 63 ♀—*a*, spider, mag; *b*, natural size; *c*, profile; *d*, cephalothorax, underside; *e*, eyes; *f*, epigyne; *g*, ♂ right palpus, from outside.

♀ Total length, 4; cephalothorax, about 1 1/6; breadth, about 2/3; breadth in front, about 1/3; abdomen, almost 3; breadth, 1 ½ millim. Leg, i.—11; ii.—7 ¼; iii.—4 ¼; iv.—7 millim. Patella + tibia, iv.—2 millim. (Thorell.)

♂ Total length, 4; cephalothorax, 2; breadth, ¾; abdomen, 2; breadth, 1 millim. Leg, i.—11 ½; ii.—8; iii.—4; iv.—6 ½ millim. Patella + tibia iv.—1 ½ millim.

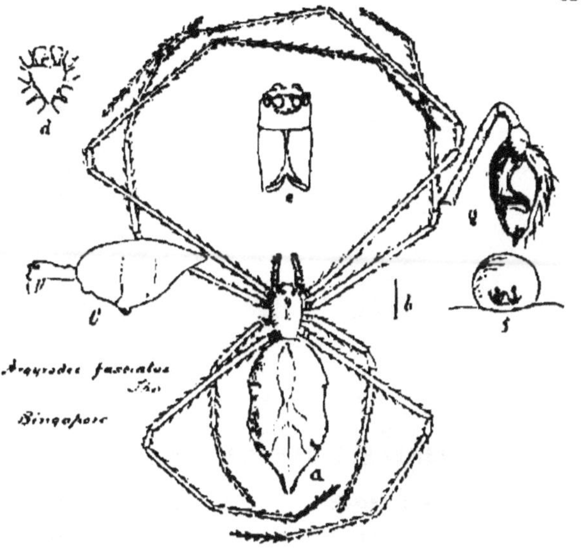
Argyrodes fasciatus
Thor.
Singapore

LINYPHIA JAVANA. sp. n.

Description of Plate 64 ♂—*a*, spider, mag.; *b*, natural size; *c*, profile; *d*, cephalothorax, underside; *e*, eyes; *f*, right palpus, from outside; *g*, do. in front.

Total length, 3; breadth, about ¾; abdomen, 1½; breadth, about 1⅓ millim. Leg, i.—6; ii.—6; iii.—4; iv.—? millim.

Type in my collection.

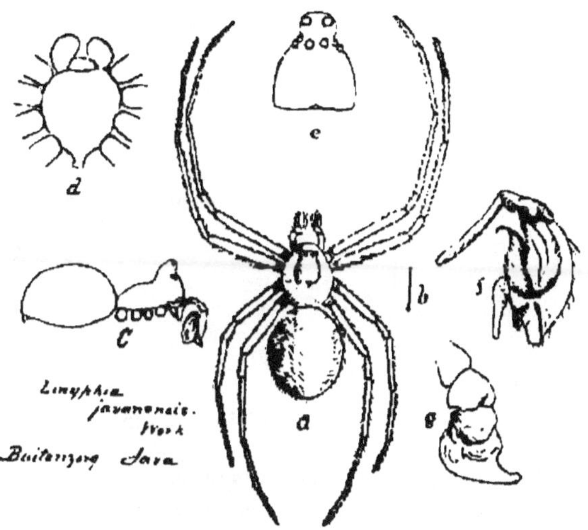
e
d
b
f
C
Linyphia
javanensis.
Work
Buitenzorg Java
a
g

THERIDIUM T-NOTATUM. Thor.

1895. *Theridium T-notatum*, Thor. The Spiders of Burma, p. 97.

Description of Plate 65 ♀ —*a*, spider, mag. ; *b*, natural size ; *c*, profile ; *d*, cephalothorax ; *e*, eyes ; *f*, epigyne.

Total length, $3^1/_3$; cephalothorax, less than 1½ ; breadth, nearly 1 ; breadth in front, scarcely $^1/_3$; abdomen, 2½ ; breadth, 2 millim. Leg, i.—scarcely 6 ; ii.—4 ; iii.—about 3 ; iv.—4¼ millim. Patella + tibia, iv.—less than 1¾ millim. (Thorell.)

T. T-notatum Thor. makes a small reticulated snare, and sits with its cocoon under a twisted leaf in centre.

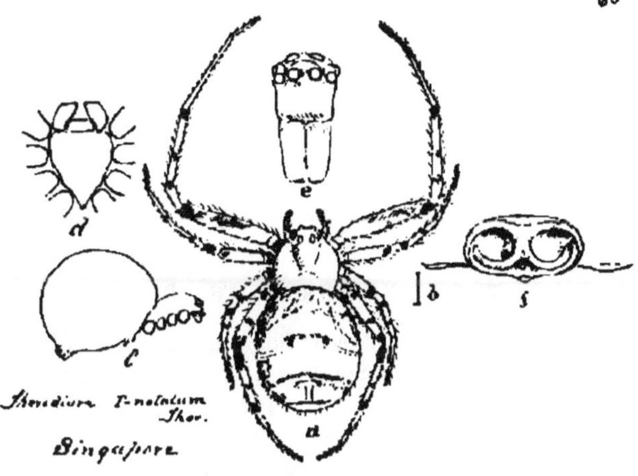
e
d
b
f
c
a
Theridium T-notatum Thor.
Singapore

THERIDIUM TUBICOLUM. Dol.

1859. *Theridium tubicolum.* Dol. Tweede Bijdr., etc., p. 49, pl. VII, figg. 7 and 7b.
1878. „ „ Thor. Studi., etc., ii., pp. 158 and 300.
1881. „ „ id. ibid. iii., p. 172.
1890. „ „ id. ibid. iv., vol. I., p. 273.

Description of Plate 66 ♀—*a*, spider, mag.; *b*, natural size; *c*, profile; *d*, cephalothorax, underside; *e*, eyes; *f*, epigyne; *g*, cocoon, mag.

Total length, 8¼; cephalothorax, 3⅙; breadth, scarcely 2¾; breadth in front scarcely 1; abdomen, 6; breadth, 5⅓ millim. Leg, i.—17 (tibia, 3⅔); ii.—12¼; iii.—8½; iv.—13½ millim. Patella + tibia, iv.—4¼ millim (Thorell).

Doleschall says:—"This spider is common in Amboina in sunny places throughout the year on hedges and low wild bushes, and that it spins a large irregular net fastened all round. In the middle the spider makes a long perpendicular pipe-shaped sack, which is closed above and funnel-shaped at the lower edge. The spider inhabits the lower end of this sack, however, drawing nimbly into it from any danger from above."

The only specimen which I found was inhabiting a large reticulated net 12 inches high, with a twisted leaf in centre, under which the spider was sitting with its cocoon. Cocoon, ¼ inch in diameter; colour, brown.

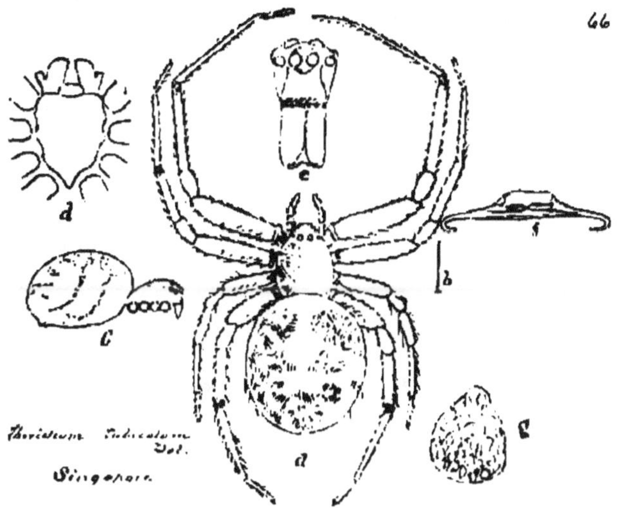
Dol.
Singapore

THERIDIUM INQUINATUM. Thor.

1878. *Theridium inquinatum*, Thor. Studi., etc., ii., p. 155.

1881. „ „ id. ibid. iii., p. 170.

1895. „ „ id. The Spiders of Burma, p. 93.

Description of Plate 67 ♀—*a*, spider, mag; *b*, natural size; *c*, profile; *d*, cephalothorax, underside; *e*, eyes; *f*, epigyne; *g*, snare, reduced.

Total length, 3¼; cephalothorax, 1¼; breadth, scarcely 1; abdomen, 2; breadth, 2 millim. Leg, i.—7; ii.—5¾; iii.—4¼; iv.—4½ millim. Patella + tibia, iv.—scarcely 1¼ millim. (Thorell.)

The snare, 3 inches in diameter, made by this spider is flat, like the groundwork of an orbicular web. Spider sits underneath a leaf suspended in the centre. No change made in snare during 24 hours. Found also in Amboina and Burma.

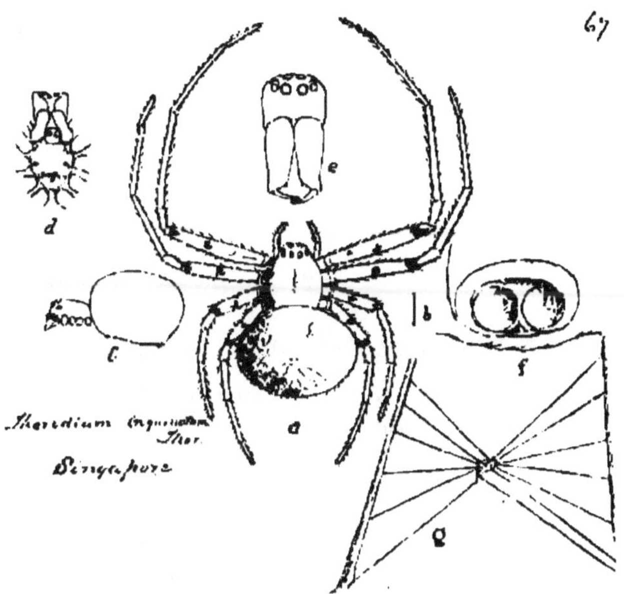
d
e
b
c
f
a
g
Theridium
Thor.
Singapore

THERIDIUM MUNDULUM, L. Koch.

1872. *Theridium mundulum*, L. Koch, Die Arachn. Austral. p. 263, pl. xxii., fig. 3.
1877. „ *amœnum*, Thor. Studi., etc., 1, p. 463 (123).
1878. „ „ id. ibid. ii., pp. 157 and 300.
1881. „ „ id. ibid. iii., p. 172.
1890 „ „ id. ibid. iv., vol. I., p. 271.
1895. „ „ id. The Spiders of Burma, p. 99.

Description of Plate 68 ♀—*a*, spider, mag; *b*, natural size; *c*, profile; *d*, cephalothorax, underside; *e*, eyes; *f*, epigyne; *g*, cocoon, mag.

Total length, 4½; cephalothorax, 1½; breadth, almost 1¼; abdomen, almost 3½; breadth, 2²/₃; height, almost 3 millim. Leg, i.—7¹/₅ (tibia 1⅔); ii.—5; iii.—3²/₃; iv.—5¾ millim. Patella+tibia iv.—2 millim. (Thorell.)

This spider, which is common around Singapore, makes a reticulated snare, in the centre of which it lives under a twisted leaf. Its cocoon, which is yellow, is also placed under the leaf. Found also in Amboina, Celebes, and Burma.

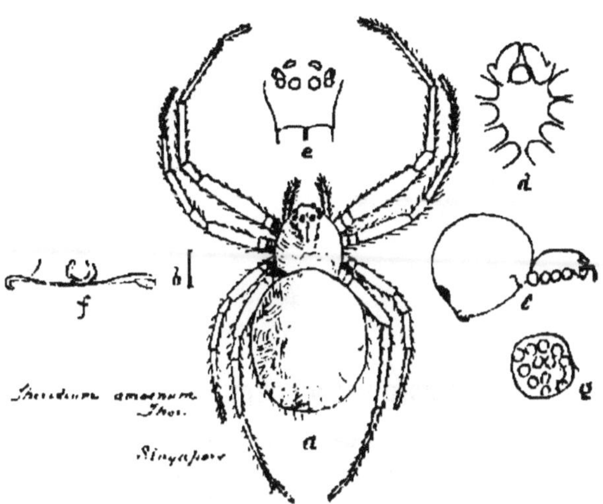
e
d
b
f
c
g
Theridium amoenum
Thor.
a
Singapore

THERIDIUM WEBERI. Thor.

1892. *Theridium weberi*, Thor., Novae Species Aranearum, etc., p. 11.

Description of Plate 69 ♂—*a*. spider, mag.; *b*, natural size; *c*, profile; *d*, cephalothorax, underside; *e*, eyes; *f*, epigyne; *g*, do. profile.

Total length, 4; cephalothorax, not quite 2; breadth, about 1²/₃; abdomen, scarcely 3; breadth, 2¼ millim. Leg, i.—9½; ii.—7; iii.—scarcely 5¼; iv.—7½ millim. Patella + tibia iv.—2½ millim. (Thorell.)

Was found under a rotten piece of bark on a branch, at the one side of which it had its small reticulated web.

Type in my collection.

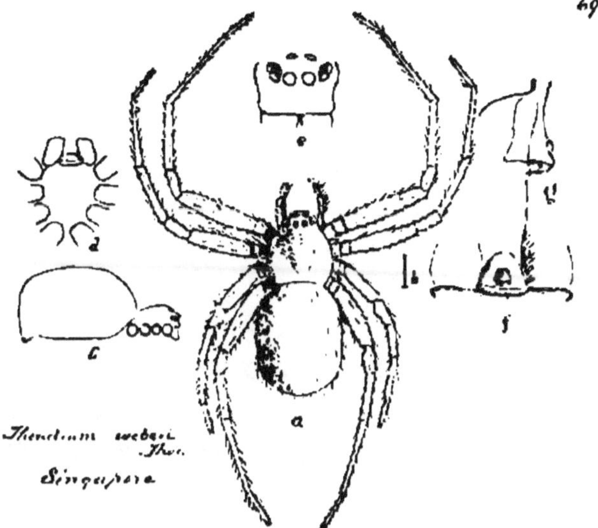
e
d
c
a
b
b'
f
Theridium weberi
Thor.
Singapore

THERIDIUM HELOPHORUM. Thor.

1895. *Theridium helophorum*, Thor. The Spiders of Burma, p. 98 (note).

Description of Plate 70 ♀—*a*, spider, mag.; *b*, natural size; *c*, profile; *d*, cephalothorax underside; *e*, eyes; *f*, epigyne; *g*, do. profile; *h*, snare, reduced; *i*, cocoon, mag.

Total length, 2¼; cephalothorax, about $1^1/_6$; breadth, almost 1; abdomen, about 1¾; breadth, 1½ millim. Leg, i.—almost $4^2/_3$; ii.—scarcely 3; iii.—about $2^1/_6$; iv.—about 4 millim. Patella + tibia, scarcely 1 millim. (Thorell.)

This spider makes an irregularly dome-shaped shelter, from ½ to ¾-inch in diameter, suspended among reticulated lines, beneath the summit of which it sits back downwards. I found two specimens in Java, the snare about 2 inches across, being constructed between rails. Cocoon, 5 x 2½ millim. more than twice the size of the spider.

Type in my collection.

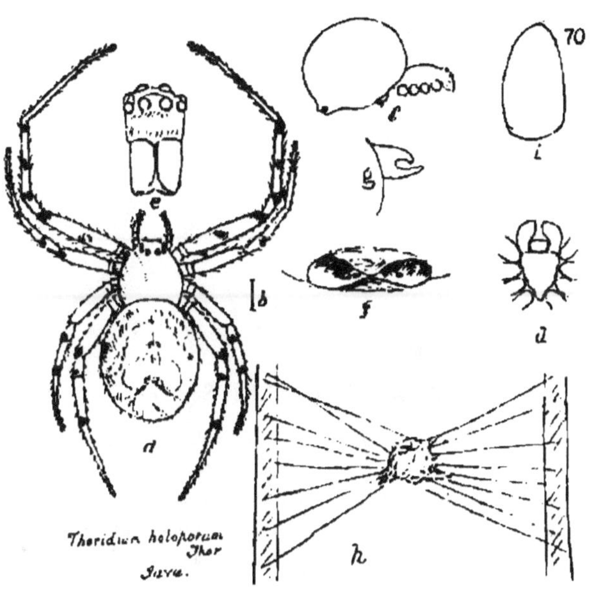
70
e
b
c
g
i
f
d
a
h
Theridium holoporum Thor
Java.

PHOLCUS GRACILLIMUS. Thor.

1890. *Pholcus gracillimus*, Thor. Studi., etc., iv., vol. I., p. 298.

Description of Plate 61 ♀—*a*, spider, mag. ; *b*, natural size ; *c*, profile ; *d*, cephalothorax, underside ; *e*, eyes ; *f*, epigyne ; *g*, ♂ right palpus from outside.

♀ Total length, 4¾ ; cephalothorax, about 1 ; breadth, about 1 ; abdomen, 3¾ ; breadth, scarcely ¾ millim. Leg, i.—(35) ; ii.—23½ ; iii.—16 ; iv.—22 millim Patella+tibia, iv.—5½ millim. (Thorell.)

♂ Total length, 5½ ; cephalothorax, 1½ ; breadth, 1 ; abdomen, 4 ; breadth, 1 millim. Leg, i.—? ; ii.—24 ; iii.—16 ; iv.— 24 millim. Patella+tibia, iv.—6 millim.

Makes a large reticulated snare under shrubs, somewhat tubular. Spider lives in the hollow centre back downwards, and carries its cocoon in its jaws. Cocoon contains 12 eggs.

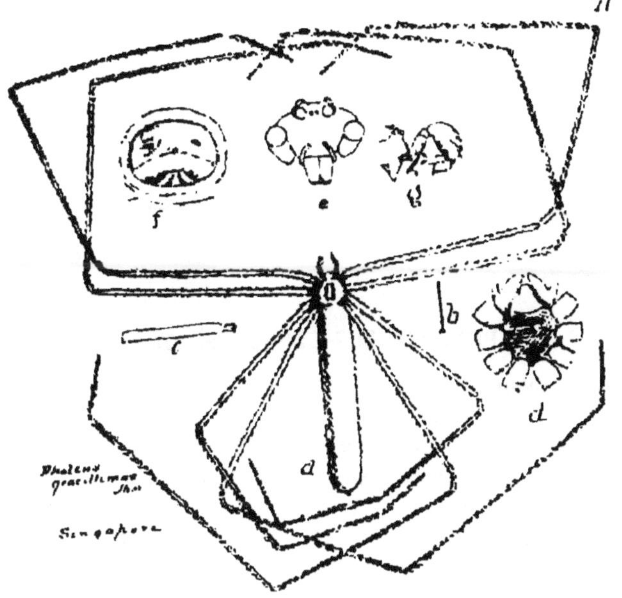
f
e
g
b
c
d
d
Singapore

PHYSOCYCLUS GLOBOSUS. Tacz.

1873. *Pholcus globosus*, Taczanowski. Les Aran. de la Gyane Francaise Horæ Societatis Ento Rossicæ, T.X. n. 1 and 2, p. 50 (Warsaw).

Description of Plate 72 ♀—*a*, spider, mag.; *b*, natural size; *c*, profile; *d*, cephalothorax, underside; *e*, epigyne and spinnerets; *f*, eyes from above; *g*, do. in front; *h*, epigyne and spinnerets in profile.

Total length, 3¾; cephalothorax, 1¾; breadth, 1¾; abdomen, 2; breadth, 2; height, 2 millim. The legs were broken off before I had opportunity of making measurements,

When a box of household effects, that I had brought home from Singapore, was opened, 21st May, 1893, at Craigdarragh, County Down, Ireland, a specimen of this spider was found living in it.

Only one specimen found by me.

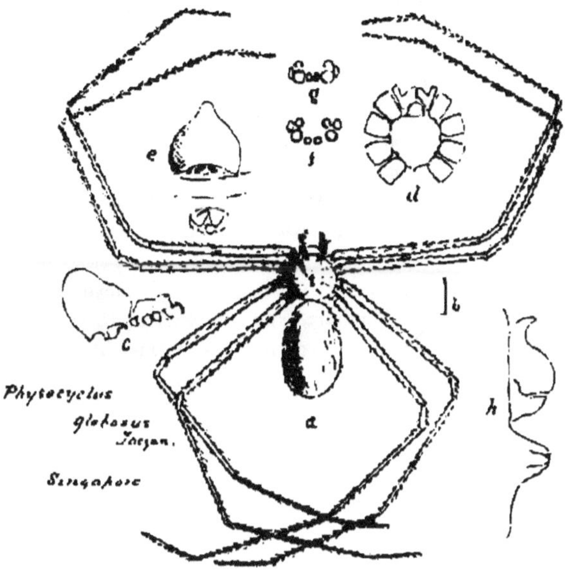
g
f
e
d
c
b
h
a
Phytocyclus
globosus
Jaczon.
Singapore

SCYTODES MARMORATA. L. Koch.

1872. *Scytodes marmorata*, L. Koch, Die Arachn. Austral. p. 292, pl. xxiv., figg. 4-4c.
1877. „ „ Thor. Studi., etc., 1, p. 471 (131).
1878. „ „ id. ibid. ii., pp. 165, 301.
1881. „ „ id. ibid. iii., p. 180.
1890 „ „ id. ibid. iv., vol. I., p. 300.
1895. „ „ id. The Spiders of Burma, p. 66.

Description of Plate 73 ♀—*a*, spider, mag; *b*, natural size; *c*, profile; *d*, cephalothorax, underside; *e*, eyes; *f*, epigyne.

Total length, 8½; cephalothorax, 4; breadth, 3; abdomen, 4½ millim. Leg, i.—33½; ii.—26; iii.—18; iv.—24¾ millim. Length of falces, 5/6 millim. (Thorell.)

Prof. Thorell believes that *S. thoracica* Var. *indica* Van Hass. Aran. exot., 1871, p. 5, is to be referred to this species.

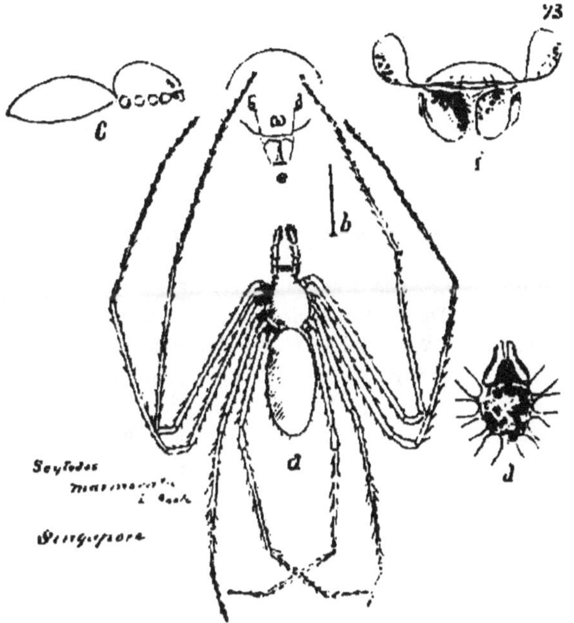
73
c
i
b
a
d
Scytodes
marmorata
L. Koch
Singapore

DESIS MARTENSII. L. Koch.

1872. *Desis martensii*, L. Koch, Die Arachn. Austral., I., p. 347, pl. xxix., figg. 2-29.
1894. „ „ Thor. Arachnider fran Java, etc., p. 5.

Description of Plate 74 ♀—*a*, spider, mag.; *b*, natural size; *c*, profile; *d*, cephalothorax underside; *e*, eyes; *f*, epigyne; *g*, ♂ right palpus; *h*, nest in coral in disused lithodomus burrow.

♀ Total length, 12; cephalothorax, 6; breadth, 4; do. in front, 2; abdomen, 6; breadth, 3¾ millim. Leg, i.—18; ii.—15½; iii.—13½; iv.—18 millim. Patella+tibia, iv.—6 millim.

♂ Total length, 8; cephalothorax, 5; breadth, 2; do. in front, ½; abdomen, 4½; breadth, 2½ millim. Leg, i.—15; ii.—10; iii.—9½; iv.—13 millim. Patella+tibia, iv.—4 millim.

This spider was found by me on the Blacka Mati coral reef off the New Harbour, Singapore, the place where it was first discovered by Dr. Martens in 1861. Dr. L. Koch points out in his Arachniden Australiens, in a very interesting way, page 350, that this spider, though living on a coral reef, is not by any means a water spider, and accurately surmises the way in which it lives. I found that it was perfectly helpless when placed in a bottle of water, showing in every way that it was not in its natural element. It lives in holes made by a species of lithodomus, and spins a matted web across the hole, and so keeping an air chamber for itself during flood tide. It is found in considerable numbers, but as it runs with great rapidity, is very hard to catch.

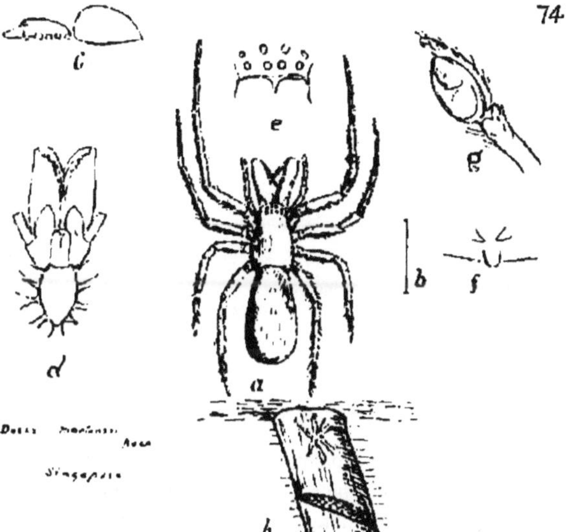

c
e
g
d
b
f
a
h

STORENA JUVENCA. sp. n.

Description of Plate 75 ♀—*a*, spider, mag; *b*, natural size; *c*, profile; *d*, cephalothorax, underside; *e*, eyes; *f*, epigyne; *g*, ♂ left palpus from outside; *h*, do. in front.

♀ Total length, 5½; cephalothorax, 2½; breadth, 2; abdomen, 3½; breadth, 2½ millim. Leg, i.—8; ii.—6; iii.—6; iv.—9 millim. Patella+tibia, iv.—2¼ millim.

♂ Total length, 5; cephalothorax, 2½; breadth, 1¾; abdomen, 2½; breadth, 2 millim. Leg, i.—9; ii.—6½; iii.—6; iv.—8 millim. Patella+tibia, iv.—2½ millim.

This spider was found under leaves.

Type specimens 2 ♂ and 1 ♀ in my collection.

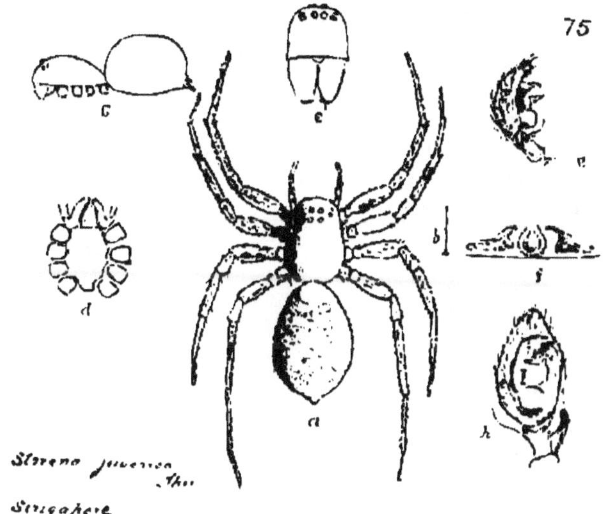
75
c
e
g
d
b
f
a
h
Storena javanica
Thor.
Singapore

STORENA ANNULIPES. Thor.

1892. *Storena annulipes*, Thor., Novae Species Aranearum, etc., p. 1.

Description of Plate 76 ♀—*a*, spider, mag.; *b*, natural size; *c*, profile; *d*, cephalothorax, underside; *e*, eyes; *f*, epigyne.

Total length, 8; cephalothorax, 4; breadth, 3½; abdomen, 5; breadth, 3 millim. Leg, i.—10½; ii.—11½; iii.—11; iv.—13 millim. Patella+tibia, iv.—3 millim.

Found running on the ground.

Type in my collection.

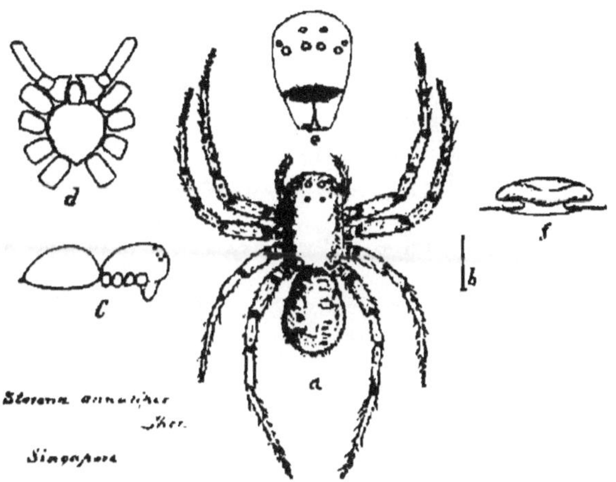
d
e
c
a
b
f
Singapore

CRYPTOTHELE SUNDAICA. Thor.

1890. *Cryptothele sundaica*, Thor., Arach. di Pinang, etc., p. 305 (41).
1894. " " id., Arachnider fran Java, etc., p. 9.

Description of Plate 77 ♀ —*a*, spider, mag.; *b*, natural size; *c*, profile; *d*, cephalothorax, underside; *e*, eyes; *f*, epigyne.

Total length, 6½; cephalothorax, 3¼; breadth, 2½; abdomen, 3½; breadth, 3½ millim. Leg, i.—7½; ii.—7; iii.—6½; iv.—7 millim. Patella+tibia, iv.—2 millim.

This spider was found under leaves, and was very sluggish in its movements.

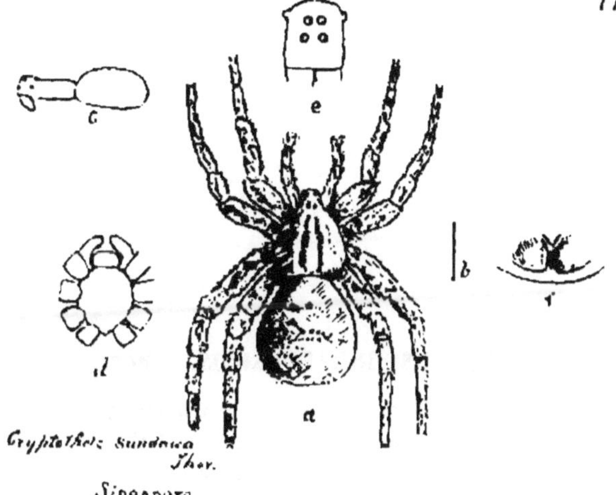
e
c
d
b
a
Cryptothele sundaica
Thor.
Singapore

PSECHRUS SINGAPORENSIS. Thor.

1894. *Psechrus singaporensis*, Thor., Decas, Aranearum, etc., p. 4.

Description of Plate 78 ♀ —*a*. spider, mag.; *b*, natural size; *c*, profile; *d*, cephalothorax, underside; *e*, eyes; *f*, epigyne; *g*, snare, reduced.

Total length, 11; cephalothorax, $4^2/_3$; breadth, 3¼; breadth of clypeus, 1½; abdomen, 7; breadth, 2 millim. Leg, i.—$38^1/_3$; ii.—27¾; iii.—19½; iv.—$30^2/_3$ millim. Patella+tibia iv.—9 millim. (Thorell.)

This spider make a large reticulated snare, about two feet long, in banks or among stones, and generally takes advantage of some hole or crevice, to which one end of the net is fastened, to form a retreat in time of danger. It remains generally in the middle of the snare, but is easily frightened, and runs with great rapidity into its den. Sparsely found in Singapore.

Type in my collection.

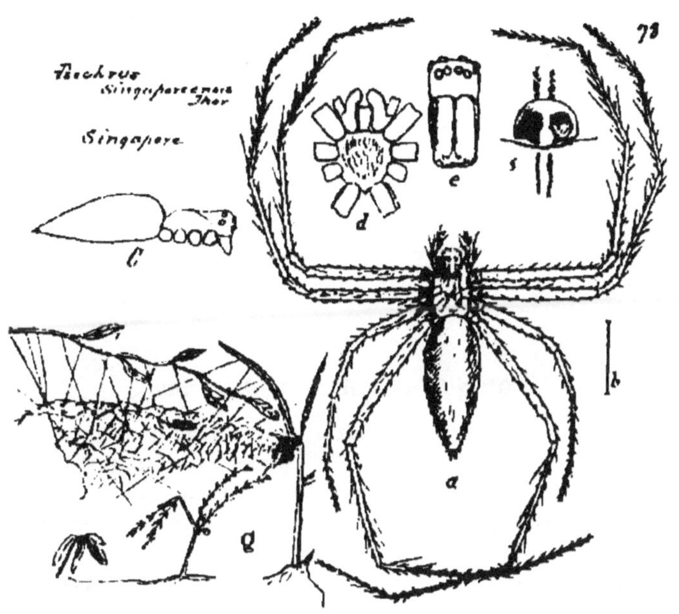
78
Thor
Singapore
a
b
C
d
e
f
g

CORINNOMMA HARMANDII. Sim.

1886. *Corinnomma harmandii*, Sim. Arachn. recueillis, etc., p. 24 (= ♀).

1887. " " Thor. Ragni Birmani, etc., p. 45 (= ♂).

1895. " " id. The Spiders of Burma, p. 40.

Description of Plate 79 ♀—*a*, spider, mag; *b*, natural size; *c*, profile; *d*, cephalothorax, underside; *e*, eyes; *f*, epigyne; *g* do., profile.

Total length, 9; cephalothorax, 5; breadth, 2¾; abdomen, 4½; breadth, 3½ millim. Leg, i. and ii.—about 10½; iii.—?; iv.—14 millim. Patella + tibia, iv.—4 millim.

My note in regard to the only specimen found by me is—"Found in a twisted up leaf in a reticulated snare, but I think it is not the fabricator of the snare."

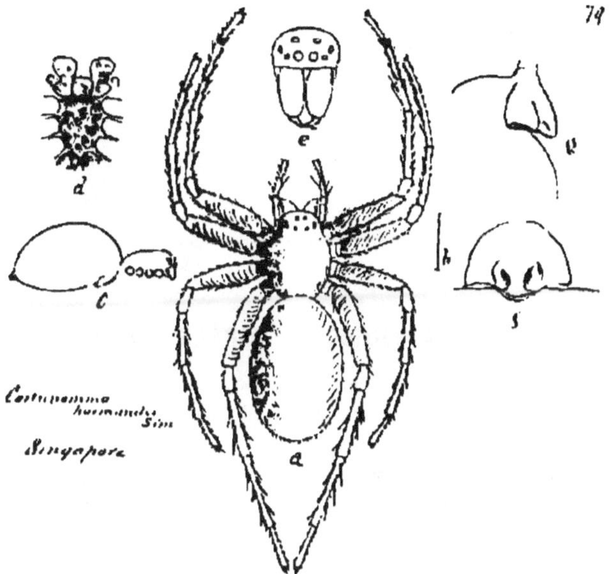
79
d
e
u
c
h
s
a
Sim
Singapore

CHIRACANTHIUM MANGIFERÆ. sp. n.

Description of Plate 80 ♀—*a*, spider, mag.; *b*, natural size; *c*, profile; *d*, cephalothorax, underside; *e*, eyes; *f*, epigyne; *g*, ♂ right palpus, from outside; *h*, do. from below.

♀ Total length, 8½; cephalothorax, 4; breadth, 2¼; do. in front, 1½; abdomen, 5½; breadth, 2 millim. Leg, i.—9; ii.—10; iii.—6½; iv.—12 millim. Patella+tibia, iv.—4 millim.

♂ Total length, 7; cephalothorax, 3; breadth, 2½; abdomen, 4; breadth, 2 millim. Leg, i.—22; ii.—15; iii.—10; iv.—16 millim. Patella+tibia, iv.—5 millim.

C. mangiferæ constructs on the face of a mango leaf a closely woven curtain of white silk, about 1 inch long by ½ an inch wide, behind which it lives.

Its cocoon, made of eggs slightly matted together, was also fastened under this covering.

Type in my collection.

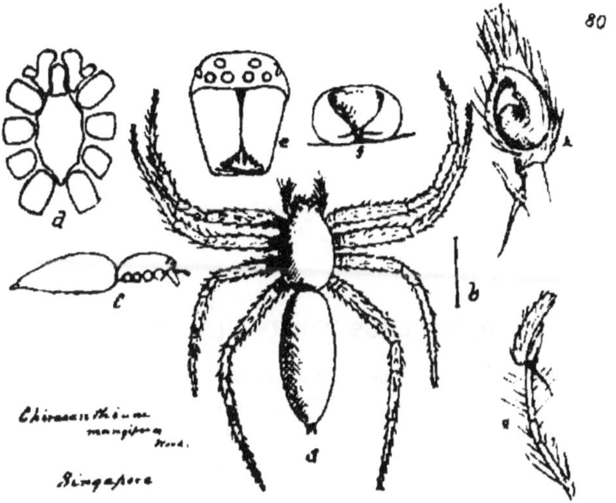

80
a
e
b
c
Chiracanthium mangiferae Work.
Singapore

DAMARCHUS WORKMANII. Thor.

1891. *Damarchus workmanii*, Thor. Spindlar fran Nikobarerna, etc., p. 15.

1895. „ „ id. The Spiders of Burma, p. 4.

Description of Plate 81 ♀—*a*, spider, mag. ; *b*, profile ; *c*, eyes ; *d*, cephalothorax, underside ; *e*, natural size of spider ; *f*, nest, reduced ; 1, 2, 3, 4 terminal claws on 1, 2, 3, and 4 pair of legs.

Total length, 19½ ; cephalothorax, 8½ ; breadth, almost 6 ; breadth of clypeus, about 4 ; abdomen, 11¹/₃ ; breadth, almost 7 millim. Palpi, 13½. Leg, i.—23½ ; ii.—20½ ; iii.—17½ ; iv.—25 millim. Patella+tibia, i.—8 ; do. iv.—8 millim. Length of falces, 5 ; mamillae sup. 4 millim. (Thorell.)

D. workmanii, Thor. makes a simple tube in the ground about 3 inches deep, without any trap door, and without raising the tube above the ground or in any way protecting it. In one case I found a lateral chamber, opening slightly larger than the tube, also without a door. ♂ unknown.

Type in my collection.

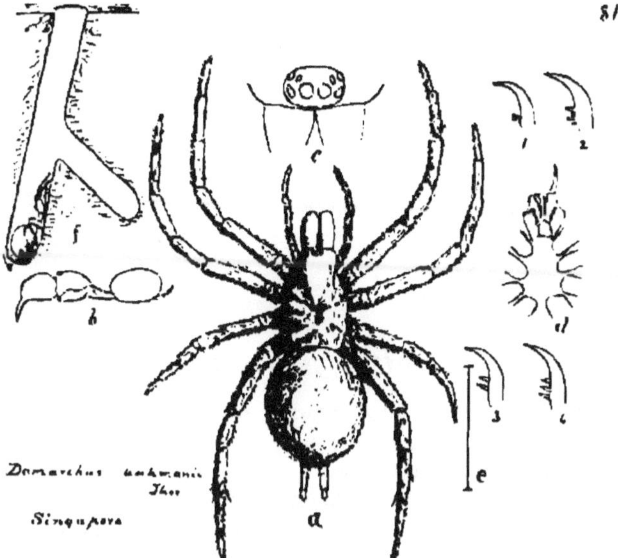
a
b
c
d
e
f
1
2
3
4
Damarchus
Thor
Singapora

SAROTES CURSOR. Thor.

1894. *Sarotes cursor*, Thor. Decas Aranearum, etc., p 19.

Description of Plate 82 ♀—*a*, spider, mag.; *b*, natural size; *c*, profile; *d*, cephalothorax, underside; *e*, eyes; *f*, epigyne (wet).

Total length, 12; cephalothorax, almost 5½; breadth, 5½; breadth of clypeus, 3½; abdomen, 7½; breadth, almost 4 millim. Leg, i.—28; ii.—28; iii.—?; iv.—23¼ millim. Patella + tibia, iv.—about 7¹/₆ millim. (Thorell).

This spider was living with its young ones (about 50 in number) in a twisted leaf. The leaf was so twisted up and fastened that one could not see either spider or cocoon.

Cocoon of white silk, 2 inches by ½ inch broad, and attached to the leaf.

Type in my collection.

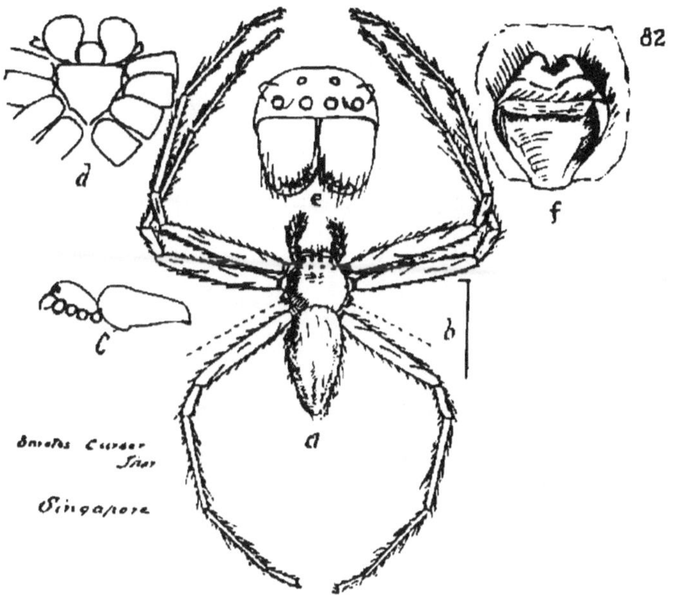
82
d
e
f
c
b
a
Singapore

LOXOBATES ORNATUS. Thor.

1891. *Loxobates ornatus*, Thor. Spindlar fran Nikobarerna, etc., p. 89.

Description of Plate 83 ♀—*a*, spider, mag; *b*, natural size; *c*, profile; *d*, maxillae and labium; *e*, eyes; *f*, epigyne.

Total length, 6; cephalothorax, 2¼; breadth, 2; breadth of clypeus, a little more than 1½; abdomen, 4; breadth, 1⁴/₅; height (in front), a little more than 2 millim. Leg, i.—7; ii.—6¾; iii.—about 3¾; iv.—a little more than 4 millim. Patella + tibia, iv. —1½ millim. (Thorell).

Type in my collection.

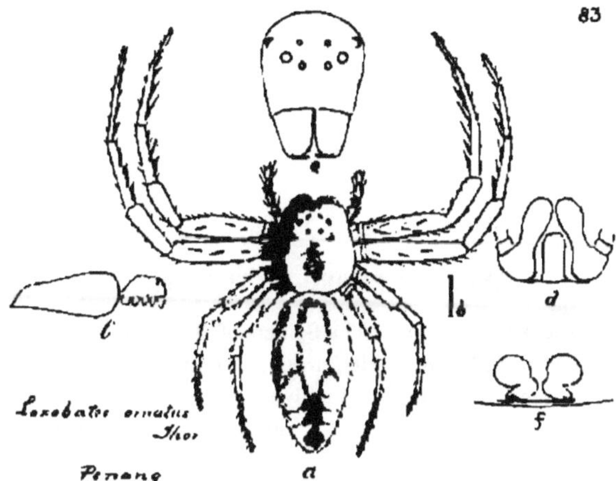
e
b
c
d
f
Loxobates ornatus
Thor
Penang
a

LYCOPUS RUBRO-PICTUS. sp. n.

1895. *Lycopus rubro-pictus*, Thor. The Spiders of Burma, p. 286.

Description of Plate 84 ♀—*a.* spider, mag.; *b*, natural size; *c*, profile; *d*, cephalothorax, underside; *e*, eyes; *f*, epigyne (hardly mature).

Total length, 6; cephalothorax, 3½; breadth, 3; abdomen, 4½; breadth, 2 millim.

Unfortunately this spider after it was drawn became denuded of all its legs, so I cannot give the measurements, but the drawing of them may be taken as fairly correct.

Cephalothorax orange coloured, slightly tinged with green, clypeus edged with red, lateral eyes placed on greyish blue tubercles, eye area large; falces yellowish green, legs greenish yellow speckled with red; abdomen more than double as long as broad, with the posterior half gradually tapering into a kind of tail, yellowish with the upper part densely speckled with red, and moreover marked with two small red spots somewhat before the middle, and with five or six transverse slightly undulated red lines on the posterior tail-like portion.

When living this spider was of a bright grass green, and had made its nest by twisting up the end of a croton leaf; in the nest was also its cocoon, out of which the young had just been hatched.

Type in my collection.

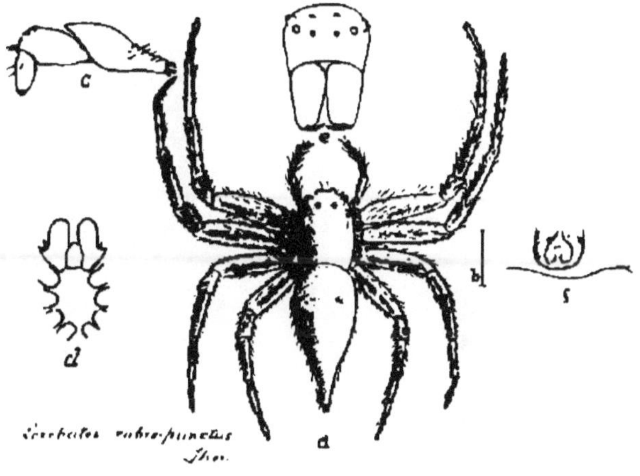

[illegible] rubro-punctus Thor.

Singapore

TMARUS PULCHRIPES. Thor.

1894. *Tmarus pulchripes*, Thor. Decas Aranearum, etc., p. 22.

Description of Plate 85 ♂—*a*, spider, mag; *b*, natural size; *c*, profile, *d*, cephalothorax, underside; *e*, eyes; *f*, right palpus, from below; *g*, do., from outside.

Total length, 3½; cephalothorax, 1⅘; breadth, a little more than 1½; breadth of clypeus, about 1; breadth in front, a little more than 1; abdomen, a little more than 2; breadth, about 1 millim. Leg, i.—almost 8; ii.—8; iii.—4; iv.—4⅔ millim. Patella+tibia, iv.—1½ millim. (Thorell.)

Type in my collection.

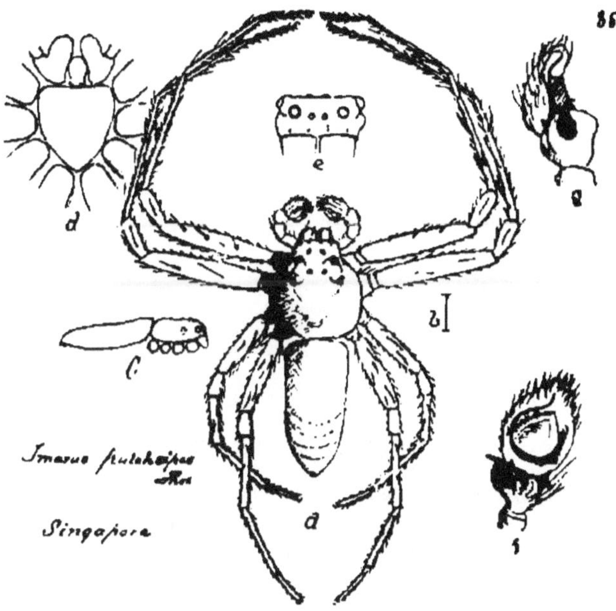
85
d
e
g
c
b
f
a
Tmarus pulchripes Thor.
Singapore

DARADIUS CALLIDUS. Thor.

1890. *Daradius callidus*, Thor. Arachnidi di Nias, etc., p. 61.

Description of Plate 86 ♀—*a*, spider, mag.; *b*, natural size; *c*, profile; *d*, cephalothorax, underside; *e*, eyes; *f*, epigyne; *g*, 1st leg showing spines.

Total length, 8½; cephalothorax, 4; breadth, slightly more than 3½; breadth in front (round bend), at least 2½; abdomen, 5 1/6; breadth, a little more than 6 millim. Leg, i.—11¼; ii.—almost 11¼; iii.—6½; iv.—6¾ millim. Patella+tibia, iv.—about 2 4/5 millim. (Thorell).

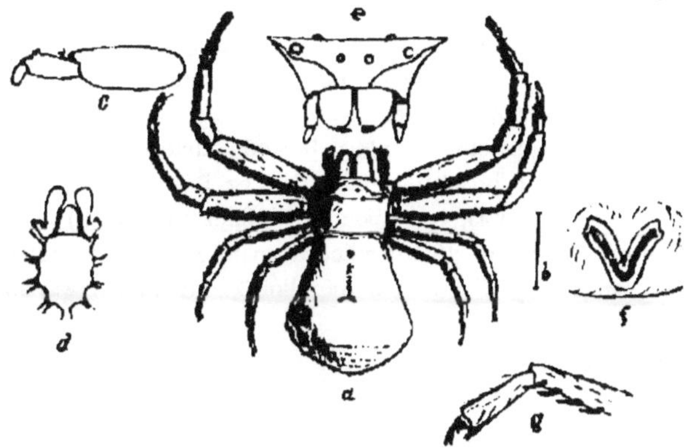

Diaeus callidus Thor
Singapore

SYNÆMA LINEATUM. Thor.

1894. *Synæma lineatum*, Thor. Decas Aranearum, etc., p. 33.

Description of Plate 87 ♀—*a*. spider, mag.; *b*, natural size; *c*, profile; *d*, cephalothorax, underside; *e*, eyes; *f*, epigyne.

Total length, 5¼; cephalothorax, 2½; breadth, almost 2½; breadth of clypeus, about 1½; abdomen, 3⅓; breadth, 3 millim. Leg, i.—7; ii.—7½; iii.—4⅔; iv.—almost 5 millim. Patella + tibia, iv.—about 1¾ millim. (Thorell.)

I have noted of this spider—"Perpendicular snare 4 inches diameter; spider sits in a little den, but cannot say if it had a trap line." Probably it was an intruder in an eperid snare.

Type in my collection.

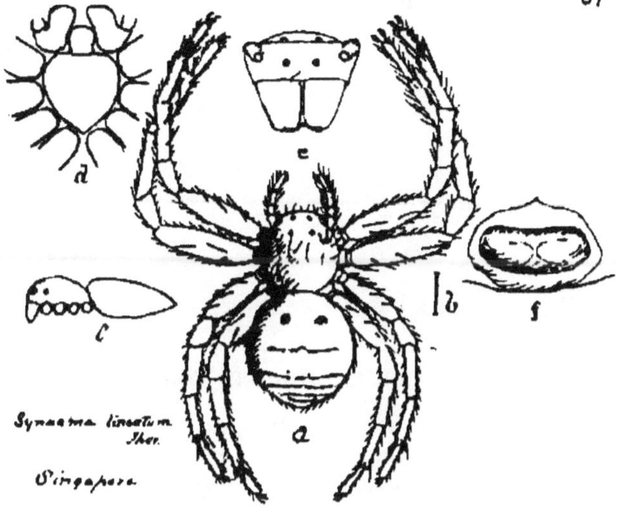
d
e
f
c
b
a
Synaema lineatum
Thor.
Singapore

ANGÆUS RHOMBIFER. Thor.

1890. *Angæus rhombifer*, Thor. Diagnoses Aran., etc., p. 19.

1892. „ „ id. Studi, etc., IV., vol. 2, p. 67.

1895. „ „ var *leucomenus*, id. The Spiders of Burma, p. 278.

Description of Plate 88 ♀—*a*, spider, mag; *b*, natural size; *c*, profile; *d*, cephalothorax, underside; *e*, eyes; *f*, epigyne (hardly mature).

Total length, 6; cephalothorax, 3; breadth, 2¾; breadth in front, 1¾; abdomen, 3; breadth, 3¼ millim. Leg, i.—9½; ii.—11¼; iii.—5; iv.—5¼ millim. Patella+tibia, iv.—2 millim.

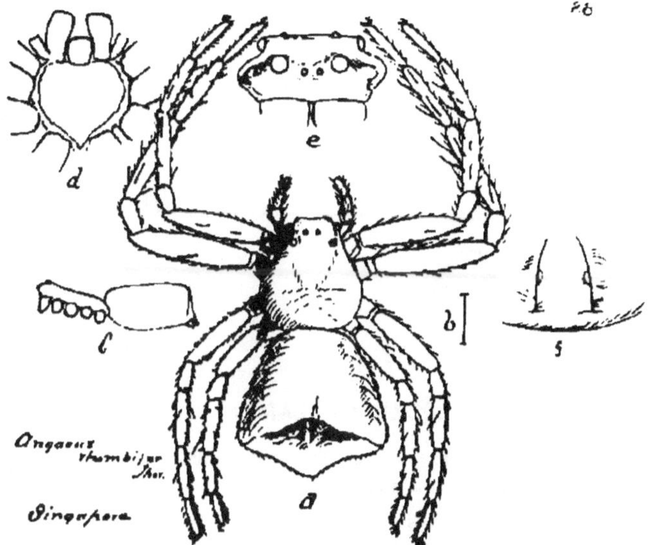
76
d
e
c
b
f
a
Angaeus rhombifer Thor.
Singapore

PHILODAMIA HILARIS. Thor.

1894. *Philodamia hilaris*, Thor. Decas, Aranearum, etc., p. 27.

Description of Plate 89 ♀—*a*, spider, mag.; *b*, natural size; *c*, profile; *d*, cephalothorax, underside; *e*, eyes; *f*, epigyne.

Total length, 3½; cephalothorax, about 1²/₃; breadth, a little more than 1½; breadth of clypeus, almost 1; abdomen, 2¼; breadth, almost 2 millim. Leg, i.—4½; ii.—almost 5; iii.—nearly 3½; iv.—3½ millim. Patella+tibia, iv.—almost 1½ millim. (Thorell.)

Type in my collection.

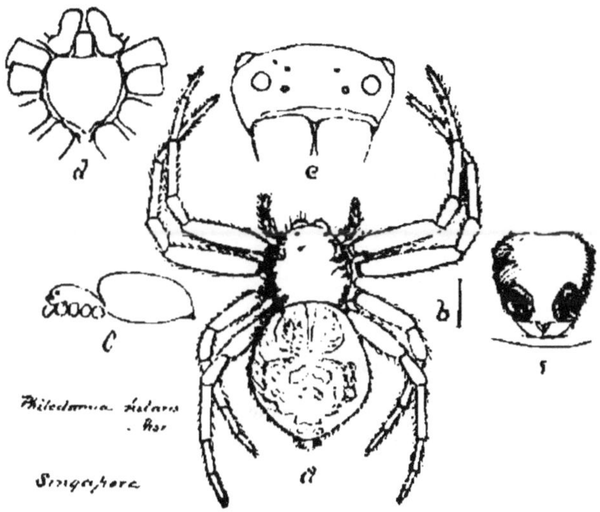
d
e
b
c
f
a
Singapore

PHILODAMIA VARIATA. Thor.

1894. *Philodamia variata*, Thor. Decas, Aranearum, etc., p. 29.

Description of Plate 90 ♀ —*a*, spider, mag. ; *b*, natural size ; *c*, profile ; *d*, cephalothorax, underside ; *e*, eyes ; *f*, epigyne.

Total length, a little more than 3½ ; length and breadth of cephalothorax, about 1¹/₃ ; breadth of clypeus, about 1 ; abdomen, 2½ ; breadth, nearly 2¹/₃ millim. Leg, i.—about 5½ ; ii.—5²/₃ ; iii.—(about 5) ; iv.—about 4¾ millim. Patella+tibia, iv.—1½ millim. (Thorell.)

Type in my collection.

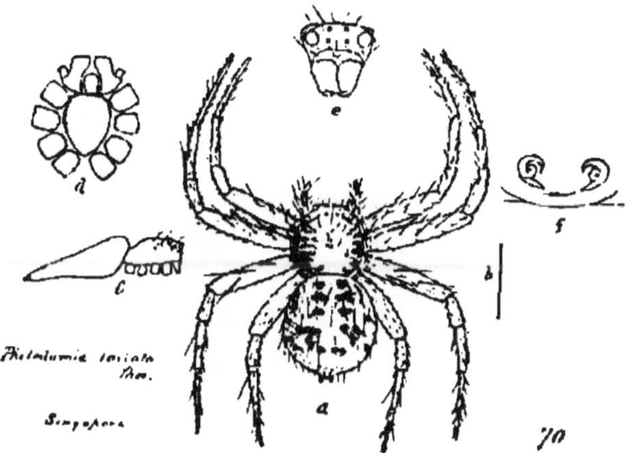
e
d
f
c
b
a
Singapore
70

DIETA VIRENS. Thor.

1891. *Orus virens*, Thor. Spindlar fran Nikobarerna etc., p. 91.

Description of Plate 91 ♀—*a*, spider, mag; *b*, natural size; *c*, profile, *d*, cephalothorax, underside; *e*, eyes; *f*, epigyne.

Total length, 10; cephalothorax, 3½; breadth, 3¼; breadth in front, 1 1/6; abdomen, 6½; breadth, 3 millim. Leg, i.—13; ii.—12½; iii.—almost 7; iv.—7 millim. Patella + tibia, iv.—2½ millim. (Thorell.)

Type in my collection.

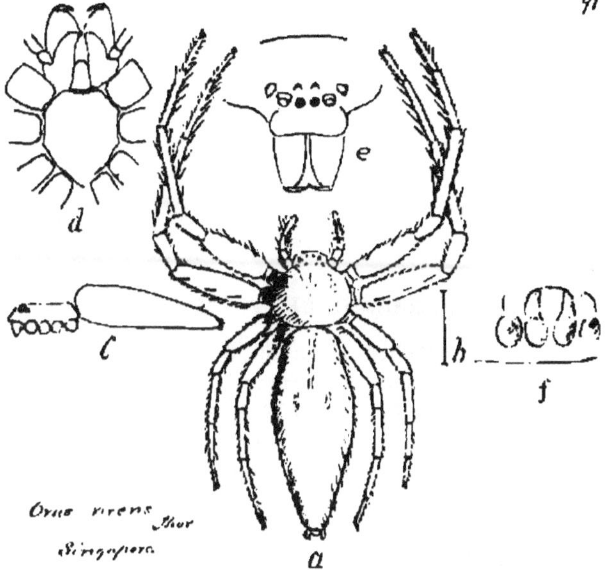

d
e
c
h
f
a
virens Thor
Singapore

PHRYNARACHNE DECIPIENS. Forbes.

1883. *Thomisus decipiens*, Forbes. On habits of *Thomisus decipiens*, p. 586, pl. 51.
1884. *Ornithoscatoides* „ Camb. On two new genera of spiders, p. 199, pl. xv., fig. 1.
1885. „ „ id. in Forbes A Naturalist's Wanderings, etc., p. 120.
1890. *Phrynarachne* „ Thor. Studi, etc., IV., vol. I, p. 35.
1890. „ „ id. Arachn. di Nias, etc., p. 94.

Description of Plate 92 ♀—*a*, spider, mag; *b*, natural size; *c*, profile; *d*, cephalothorax, underside; *e*, eyes; *f*, epigyne; *g*, spider fastened to leaf.

Total length, 14; cephalothorax, 5; breadth, 5; do. in front, 2½; abdomen, 9; breadth, 9 millim. Leg, i.—20; ii.—20; iii.—11; iv.—12 millim. Patella+tibia, iv.—4 millim.

According to Forbes, this spider is of a pure chalk white when alive, and imitates the excreta of a bird for the purpose of catching its prey.

This specimen is in my collection, and is in all probability the original type specimen described by Forbes from West Java, as this was part of his collection (noted from Java but without a number) purchased from Janion, of London.

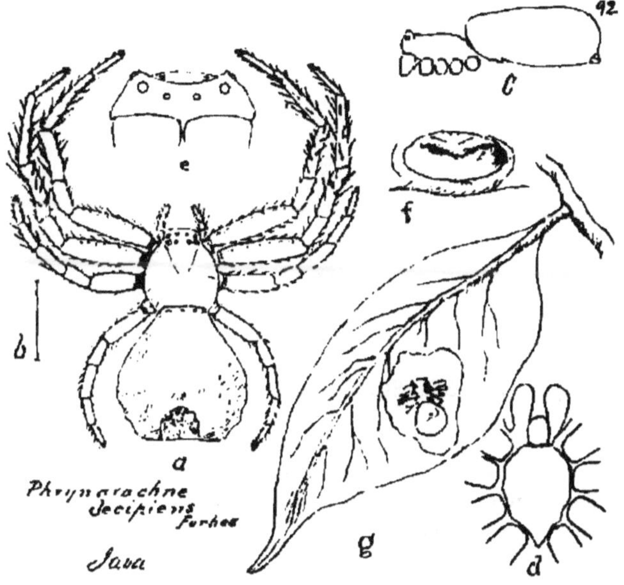
92
c
e
f
b
a
g
d
Phrynarachne
decipiens
Forbes
Java

BOLISCUS SEGNIS. Thor.

1891. *Boliscus segnis*, Thor. Spindlar fran Nikobarerna, etc., p. 98

1895. „ „ id. The Spiders of Burma, p. 283.

Description of Plate 93 ♀—*a*, spider, mag.; *b*, natural size; *c*, profile, *d*, cephalothorax, underside; *e*, eyes; *f*, epigyne.

Total length, 3½; length and breadth of cephalothorax, a little more than 1½; height, almost 1½; breadth of clypeus, 1; abdomen, 2½; breadth, 2¾; height, about 2½ millim. Legs, i. and ii.—about 2¾; iii.—about 2; iv.—about 2⅙ millim. Patella + tibia, iv.—a little more than 1 millim. (Thorell).

Type in my collection.

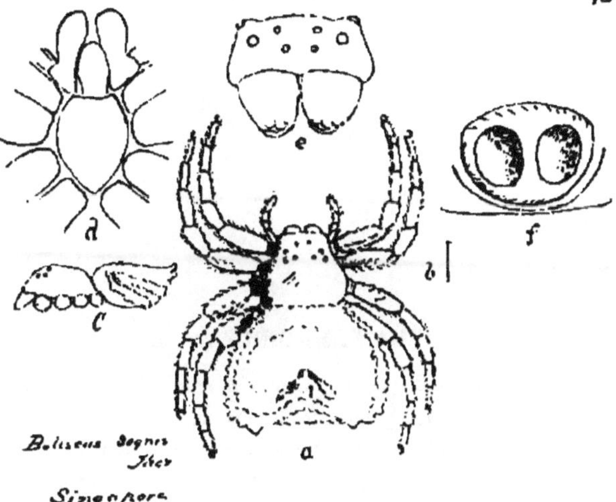
d
e
f
c
b
a
Singapore

MISUMENA SEMI-CINCTA. sp. n.

Description of Plate 94 ♂—*a*, spider, mag. ; *b*, natural size ; *c*, profile ; *d*, cephalothorax, underside ; *e*, eyes ; *f*, left palpus, from outside.

Total length, 2 ; cephalothorax, 1 ; breadth, 1 ; abdomen, 1 ; breadth, 1 millim. Leg, i.—5 ; ii.—? ; iii.—$1^2/_3$; iv.—$1^2/_3$ millim. Measurements only approximate.

Cephalothorax orange coloured, paler in the middle, with two longitudinal parallel narrow pale ferruginous bands proceeding from the lateral eyes backward, and armed with several small black spines arranged on either side of the cephalothorax (at least behind) in a few radiating rows; falces palpi and posterior legs yellowish, anterior legs darker yellowish with the patellæ, tibiæ, metatarsi, and tarsi reddish brown, abdomen yellow marked on the posterior part, with a very broad transversal black band slightly notched in the anterior margin.

Type specimen in my collection.

44

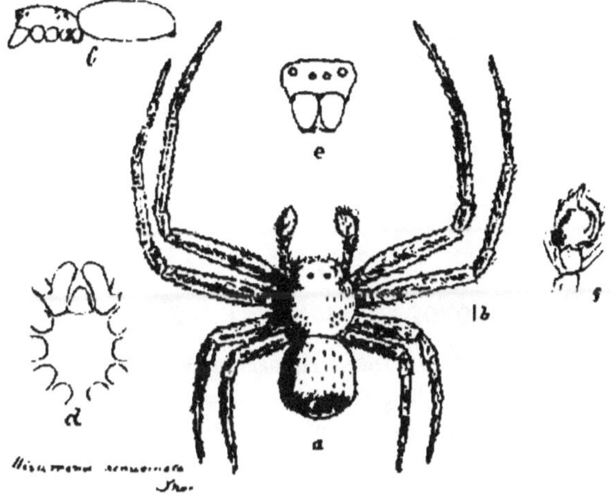

Misumena acuminata Thor.

Singapore

HYGROPODA PROGNATHA. Thor.

1894. *Hygropoda prognatha*, Thor. Decas Aranearum, etc., p. 4.

1895. „ „ id. The Spiders of Burma, p. 221.

Description of Plate 95 ♀—*a*. spider, mag.; *b*, natural size; *c*, profile; *d*, cephalothorax, underside; *e*, eyes; *f*, do., somewhat above; *g*, epigyne; *h*, ♂ left palpus, in front; *i*, do., from outside.

♀ Total length, 10½; cephalothorax, 4; breadth, about 3; breadth of clypeus, 1¾; abdomen, 6¾; breadth, 3⅓ millim. Leg, i.—29½; ii.—23½; iii.—10¼; iv.—23 millim. Metatarsus, i.—7¼; tarsus, i.—6½. Patella+tibia, iv.—6⅔ millim. (Thorell.)

♂ Total length, 9; cephalothorax, scarcely 4; breadth, scarcely 3; breadth of clypeus, about 1¾; abdomen, 5¼; breadth, 2¼ millim. Palpus, 10½; leg, i.—33; ii.—25½; iii.—12¼; iv.—24½ millim. Tibia, i.—7½; metatarsus, i.—8; tarsus, i.—7¼. Patella+tibia, iv.—7⅓ millim. (Thorell.)

This spider was found in considerable numbers on the banks of the pond in the Botanic Gardens, Singapore, where it lived under leaves, and when disturbed ran out on the surface of the water.

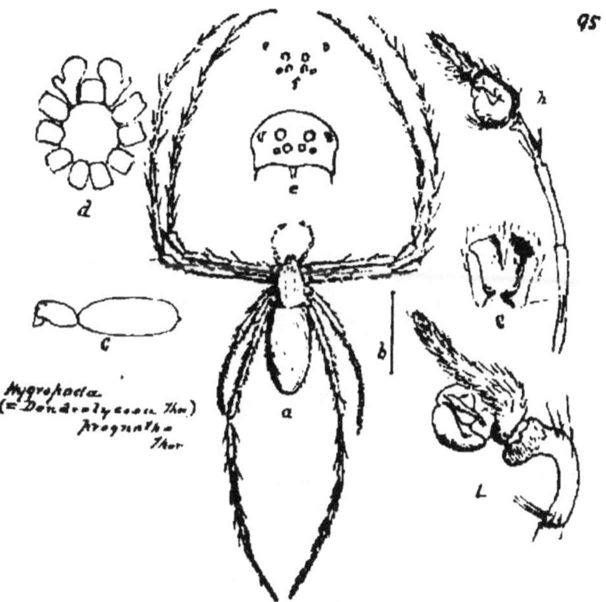
f
e
d
h
g
c
b
a
i
Hygropoda
(= Dendrolycosa Thor.)
prognatha
Thor

VENONIA CORUSCANS. Thor.

1894. *Venonia coruscans*, Thor. Decas Aranearum, etc., p. 13.

Description of Plate 96♂—*a*, spider, mag.; *b*, natural size; *c*, profile; *d*, cephalothorax, underside; *e*, eyes; *f*, palpus, from below; *g*, do, from outside; *h*, do., from inside.

Total length, 3½; cephalothorax, almost 2; breadth, a little less than 1; breadth of clypeus, nearly ½; abdomen, 2; breadth, almost 1 millim. Leg, i.—6; ii.—5¾; iii.—at least 5¾; iv.—7½ millim. Patella+tibia, iv.—about 2⅙. Metatarsus, iv.—about 2 millim. (Thorell.)

Type in my collection.

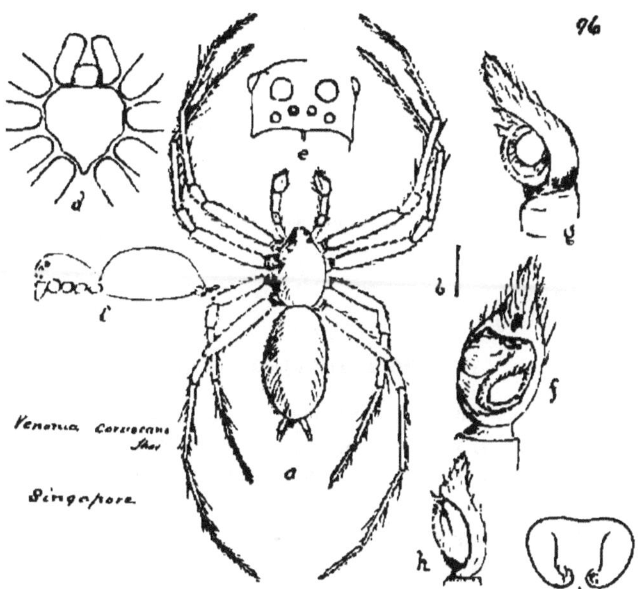
96
d
e
g
i
b
f
Venonia coruscans
Thor
Singapore
a
h
i

POLYBŒA VULPINA. Thor.

1895. *Polybœa vulpina*, Thor. The Spiders of Burma, p. 228.

Description of Plate 97 ♀—*a*, spider, mag.; *b*, natural size; *c*, profile; *d*, cephalothorax, underside; *e*, eyes, from above; *f*, do., in front; *g*, epigyne; *h*, ♂ right palpus, in front; *i*., do., from outside.

♂ Total length, 7½; cephalothorax, 4; breadth, 2½; do in front, 1½; abdomen, 4½; breadth, 2 millim. Leg, i.—19; ii.—20; iii.—17; iv.—18 millim. Patella+tibia, iv.—5½ millim.

♀ Total length, 9; cephalothorax, 4; greatest breadth, 2½; breadth in front, 1; abdomen, 5; breadth, 2¼ millim. Legs, i., ii., iv.—about 17 (ii. pair perhaps 17¼); iii.—14 millim. Patella+tibia, iv.—5 millim. Metatarsus, iv.—5 millim. (Thorell M.S.)

The ♂ and ♀ of *P. vulpina*, Thor., seem to join in constructing a large reticulated web in low shrubs, in which each sex has its own chamber about 4 inches apart.

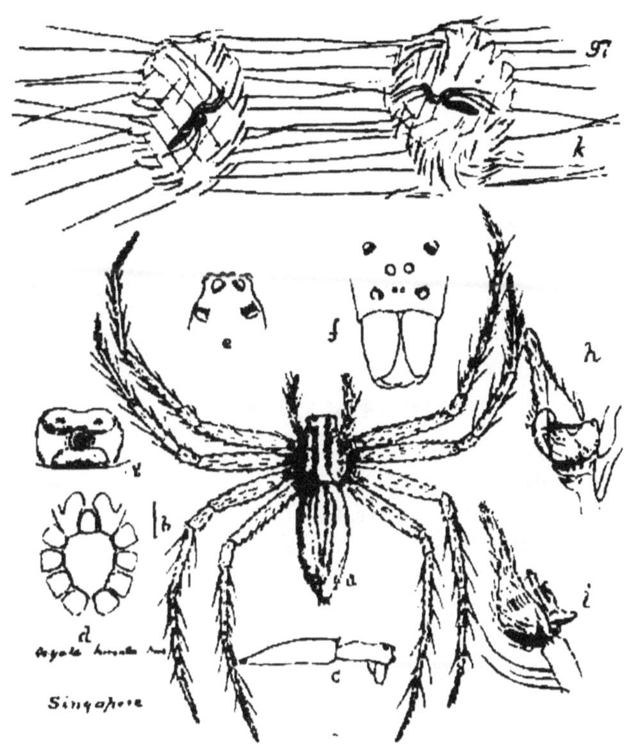
k
e
f
h
b
a
i
d
c
Singapore

TAPPONIA AUSTERA. Thor.

1894. *Tapponia austera*, Thor. Decas Aranearum, etc., p. 16.

Description of Plate 98 ♂—*a*, spider, mag.; *b*, natural size; *c*, profile, *d*, cephalothorax, underside; *e*, eyes; *f*, left palpus, from below (wet); *g*, do., from outside.

Total length, 7; cephalothorax, 3¼; breadth, a little more than 2½; breadth of clypeus, about 1 5/6; abdomen, a little more than 4; breadth, almost 2 millim. Leg, i.—12½; ii.—11¼; iii.—9½; iv.—about 9½ millim. Patella+tibia, iv.—less than 3 millim. (Thorell).

This spider was found, along with many others, in the clay nest of a dauber wasp.

Type in my collection.

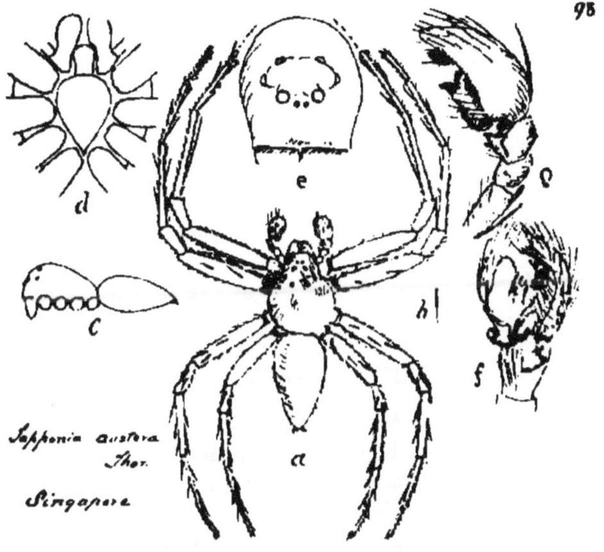
d
e
c
h
a
f
Sapponia austera
Thor.
Singapore

OXYOPES JAVANUS. Thor.

? 1886. *Oxyopes lineatipes*, Sim. Arachn. Malacca, etc., p. 16.
1887. „ *javanus*, Thor. Ragni Birmani, p. 329.
1890. „ „ id. Arachn. di Nias, etc., p. 39.
1892. „ „ id. Studi, cet., IV., 11, loc., cit., p. 195.
1894. „ „ id. Arachnider fran Java, etc., p. 11.
1895. „ „ id. The Spiders of Burma, p. 247.

Description of Plate 99 ♂—*a*. spider, mag.; *b*, natural size; *c*, profile; *d*, cephalothorax, underside; *e*, eyes; *f*, right palpus, from below; *g*, do., from outside.

Total length, 7; cephalothorax, almost 3; breadth, a little more than 2; breadth in front, about 1 1/6; abdomen, 4; breadth, 1 2/3 millim. Leg, i.—12½; ii.—11; iii.—9½; iv.—11½ millim. Patella+tibia, iv.—3¼ millim. (Thorell.)

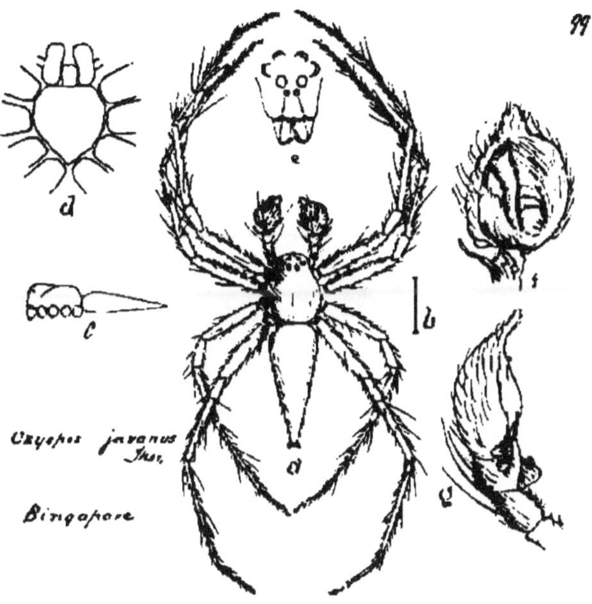
a
b
c
d
e
f
g
Oxyopes javanus Thor.
Singapore

OXYOPES BIRMANICUS. Thor.

1887. *Oxyopes birmanicus*, Thor. Ragni Birmani, etc., p. 325.
1890. „ „ id. Arachn. di Nias, etc., p. 38.
1895. „ „ id. The Spiders of Burma, p. 248.

Description of Plate 100 ♀—*a*, spider, mag; *b*, natural size; *c*, profile; *d*, cephalothorax, underside; *e*, eyes; *f*, epigyne; *g*, do. profile.

Total length, 14½; cephalothorax, almost 5; breadth, 3½; do. in front, a little more than 1½; abdomen, 9½; breadth, 4 millim. Leg, i.—22¼; ii.—20¾; iii.—16¾; iv.—20 millim. Patella+tibia, iv.—6¼ millim. (Thorell.)

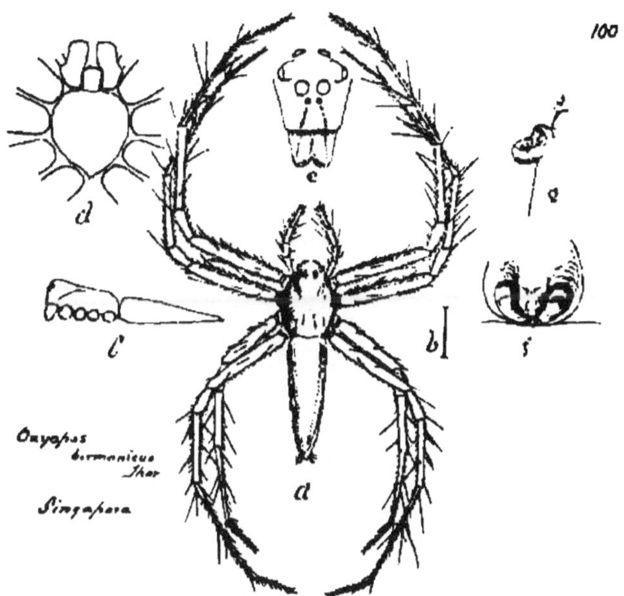

d
c
e
l'
b
i
a
Oxyopes
birmanicus
Thor
Singapora

OXYOPES AURATUS. Thor.

1890. *Oxyopes auratus*, Thor. Arachnidi di Nias, etc., p. 39.

Description of Plate 101 ♂—*a*, spider, mag.; *b*, natural size; *c*, profile; *d*, cephalothorax, underside; *e*, eyes; *f*, right palpus, from below; *g*, do. from outside.

Total length, 5; cephalothorax, almost 2½; breadth, about 1⁴/₅; breadth in front, almost 1; abdomen, 2½; breadth, a little more than 1 millim. Leg, i.—10¾; ii.—9½; iii.—8; iv.—9¾ millim. Patella+tibia, iv.—2¾ millim. (Thorell.)

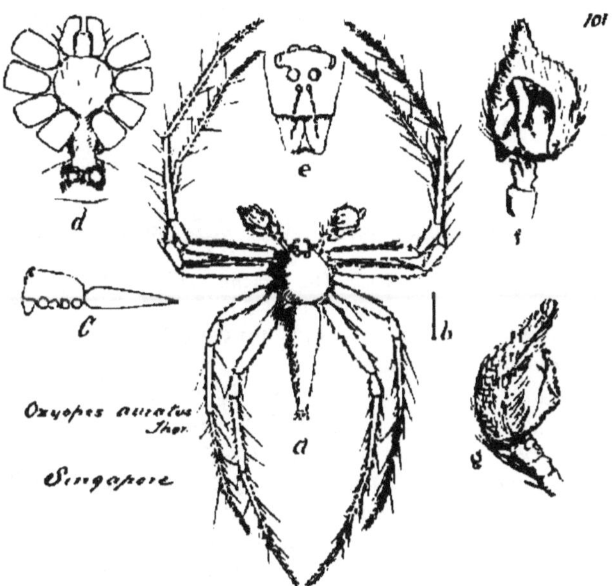
101
d
e
f
C
b
Oxyopes auratus Thor.
Singapore
a
g

HYLLUS GIGANTEUS. C. L. Koch.

1846. *Hyllus giganteus*, C. L. Koch, Die Arachn., xiii., p. 161, pl. ccclix., fig. 1216 (♂ formæ princip.)
1857. *Attus alfurus*, Dol. Bijdr., etc., p. 431 (=♂).
1857. „ *cornutus*, id. ibid. p. 432 } (=♀).
1859. „ „ id. Tweedi Bijdr., etc., pl. iv., fig. 5 } (=♀).
1859. „ „ id. ibid. p. 13, pl. xi., fig. 10 (?) } (=♀).
1859. „ *alfurus*, id. ibid. pl. iv., fig. 3 (=♂).
1877. *Hyllus giganteus*, Thor. Studi, etc., 1, p. 598 (258) (Var. *whitei*, Thor. ♂).
1878. „ „ Thor. Studi, etc., ii., p. 265.
1892. „ „ id. ibid. iv., vol. II., p. 379.

Description of Plate 102 ♀—*a*, spider, mag ; *b*, natural size ; *c*, profile , *d*, cephalothorax, underside; *e*, eyes ; *f*, epigyne.

Total length, 15¾ ; cephalothorax, 7; breadth, 6; do. in front, about 3²/₃ ; abdomen, 8¾; breadth, 5¹/₃ millim. Leg, i.—16²/₃ ; ii.—15¹/₃ ; iii.—15⁴/₅ ; iv.—17 millim. Patella + tibia, iii.—5½ ; Patella + tibia, iv.—at least 5½ ; Metatarsis + tarsus, iv.—4¾ millim. (Thorell.)

Doleschall says—"This is the largest and commonest species of this genus found in Amboina, and that their bite is very painful. He was bitten in the finger by one, and endured great pain from it, which did not seem always in the one place, for nearly seven minutes. One of his collectors, an old man, was bitten by one of them in the hand, which became immediately greatly swollen."

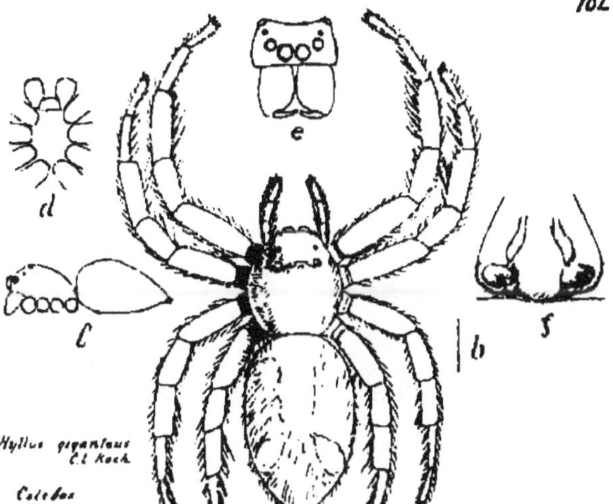
e
d
c
b
f
Hyllus giganteus
C. L. Koch
Celebes
a

VICIRIA HASSELTII. Thor.

1878. *Sinis* (?) *hasseltii*, Thor., Studi, etc., ii., pp. 6 and 274.

1879. *Viciria* „ Van Hass., Tijdschr. v. Entomol. xxii., p. 220.

1886. „ *scoparia*, Sim. Art. Soc. Linn. Bordeaux xl., p. 138.

1892. „ *hasseltii*, Thor. Studi, etc., IV., vol. 2, p. 389.

1895. „ „ id. The Spiders of Burma, p. 371.

Description of Plate 103 ♂—*a*, spider, mag.; *b*, natural size; *c*, profile; *d*, cephalothorax, underside; *e*, eyes; *f*, left palpus, from below.

Total length, 11½; cephalothorax, 5; breadth, 3⁴/₅; do. in front, 2; abdomen, 6¹/₆; breadth, a little more than 2 millim. Leg, i.—15²/₃; ii.—14½; iii.—14¾; iv.—13½ millim. Patella + tibia, i.—6; patella + tibia, iii.—4⅓; patella + tibia, iv. —4¼; metatarsus + tarsus, iv.—4¼ millim. Falce, 2¼ millim. long. (Thorell.)

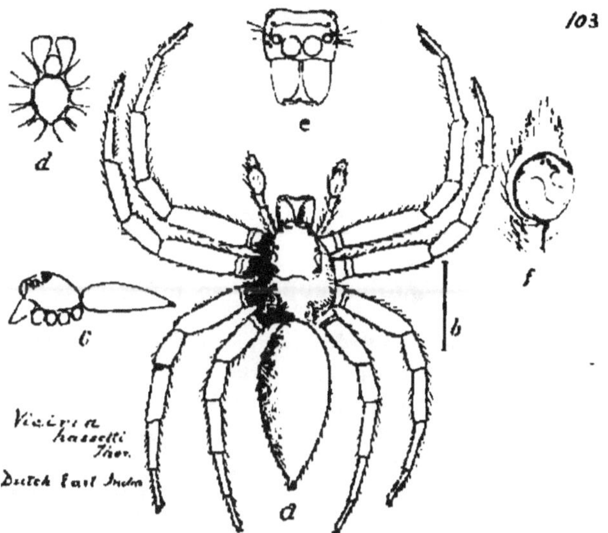

103
d
e
f
c
b
a
Viciria hasselti Thor.
Dutch East India

LAGNUS RUBER. sp. n.

Description of Plate 104 ♂—*a*, spider, mag.; *b*, natural size; *c*, profile, *d*, cephalothorax, underside; *e*, eyes; *f*, left palpus, from below (wet); *g*, do., from outside (wet).

Total length, 9: cephalothorax, 4; breadth, 2½; do. across anterior pair of eyes, 1½; abdomen, 5; breadth, 1 millim. Leg, i.—13½; ii.—12; iii.—12; iv.—10½ millim. Patella+tibia, iv.—3½ millim. Palpus, 5 millim.

This spider has considerable resemblance to *L. longimanus*, L. Koch, but differs in the shape of the falces when looked at from in front, which, in the latter species, are rounded, and in having the large front eyes close to the margin of the clypeus, but more especially in the length of the palpi. In *L. longimanus* the palpi are as long as the anterior legs.

L. ruber has a much narrower abdomen, but I think it is contracted owing to the strength of the spirit in which it is preserved.

One specimen only has been found by me. It was sitting on the lower side of a leaf with its long anterior legs stretched out in front, resembling a prawn or lobster.

Type in my collection.

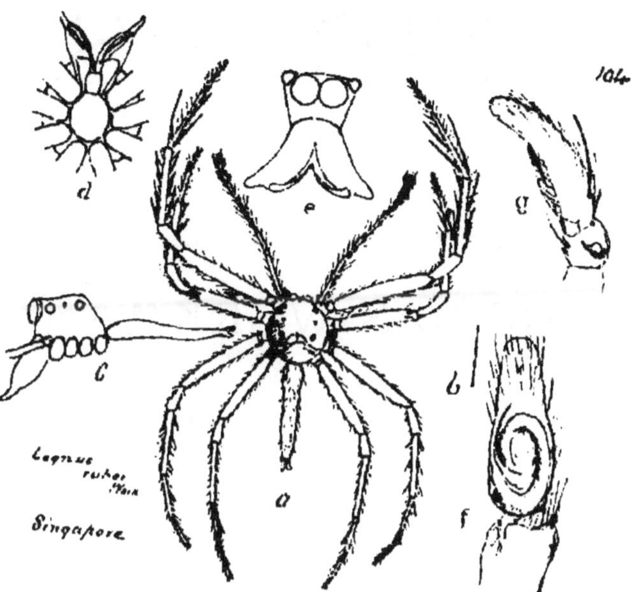
104
d
e
g
c
b
a
f
Singapore

SPIDERS.

BY

THOMAS WORKMAN.

President Belfast Natural History and Philosophical Society.

VOL. II.

MALAYSIAN SPIDERS.

PULBISHED BY THE AUTHOR.
BELFAST.
1900.

Vol. 2

CALOMATA SUNDAICA DOL.

1859. *Pelecodon sundaicus*, Dol. Tweede Bijdr, etc., p. 6, tab. XIII, figg. 2 2c.

? 1871. *Calomata sumatrana*, Auss. Beitr. Z. Kenntn. d. arachn. fam. d. Territelariae, etc. 12, 130. Taf. figg. 1 and 2.

1889-90. *do.* *sundaica*, Thor. Studi, etc. Part 4, Vol 1, p. 416.

Description of Plate 105 ♀.—*a*, spider magnified ; *b*, natural size ; *c*, cephalothorax ; *d*, profile ; *e*, eyes.

♀ Total length (without falces or spinnerets), 24½ ; cephalothorax, 9¾ ; breadth almost 7, breadth of clypeus 6 ; abdomen 14, breadth almost 10½ millim.

Leg, I.—12½, II.—13⅞, III.—13¼, IV.—14½ millim. Patella + tibia, IV.—5⅞ millim. Falce 6, falce claw 7 millim. Palpi 10½ millim. Upper spinnerets 4 millim. (Thorell).

Caphalothorax and legs of a light brown colour, abdomen dark brown.

This specimen was kindly given to me by the Curator of the Government Museum at Batavia, but nothing is known of the habits of this interesting spider.

♂ Unknown.

CHIRACANTHIUM SUAVE N.S.

Description of Plate 106 ♀.—*a*, spider magnified ; *b*, natural size ; *c*, profile ; *d*, cephalothorax underside ; *e*, eyes ; *f*, epigyne.

♀ Total length, 5½ ; cephalothorax, 2½ ; breadth, 1⅓ ; do. in front, 1 ; abdomen, 3 ; breadth, 2 ; height, 2 millim. Leg. I.—10 ; ii.—8 ; iii.—5 ; iv.—6½ millim. Patella + tibia, iv.—2 millim.

Cephalothorax and legs of a fawn colour, abdomen dark brown.

Two specimens found at Singapore. Habits unknown.

CHIRACANTHIUM CROCEUM N.S.

Description of Plate 107 ♀.—*a*, spider magnified; *b*, natural size; *c*, profile; *d*, cephalothorax underside; *e*, eyes; *f*, epigyne.

♀ Total length, 7⅓; cephalothorax, 3½; breadth, 2¼; do. in front, 1⅛; abdomen, 4; breadth, 2; height, 1½ millim. Leg, i.—7; ii.—6; iii.—5; iv.—8 millim. Patella + tibia, iv.—3 millim.

Cephalothorax and legs of a yellow colour, abdomen of a light shade.

Two specimens found at Singapore. Habits unknown.

CHIRACANTHIUM THORELLI N.S.

Description of Plate 108 ♂.—*a*, spider magnified; *b*, natural size; *c*, profile; *d*, cephalothorax underside; *e*, eyes; *f*, left palpus from outside (wet); *g*, right palpus from below (wet).

♂ Total length, 6; cephalothorax, 3; breadth, 1¾; do. in front, 1; abdomen, 3; breadth, 1¾; height, 1¾ millim. Leg, i.—11; ii.—8; iii.—6; iv.—9 millim. Patella + tibia, iv.—3 millim. Length of falce, 1¼ millim.

Similar in colour to C. croceum Work.

One specimen found at Singapore. Habits unknown.

OEDIGNATHA SCROBICULATA, THOR.

1881. *Oedignatha scrobiculata*, Thor., Studi., etc., iii., p. 209.

1889-90. *do. do. do.* Studi., etc., iv., vol. i., p. 345.

Description of Plate 109 ♂.—*a*, spider magnified; *b*, natural size; *c*, profile; *d*, cephalothorax underside; *e*, eyes; *f*, left palpus from below (wet); *g*, left palpus from outside (wet); *h*, epigyne.

♂ Total length (without falces), 4½; cephalothorax, 2; breadth, 1½; do. in front about 1; abdomen, 2; breadth, 1⅓ millim. Leg, i.—6⅓, ii.—4⅔, iii.—3, iv.—5½ millim. Patella + tibia, iv.—2 millim.

Colour rich chestnut brown.

EURYCHOERA QUDRI-MACULATA THOR.

1897. *Eurychoera quadri-maculata* Thor. Araneae paucae Asiae Australis Bihang T. K. Svenska Vet-Akad Handlingar Band 22. Afd. IV No. 6 Stockholm.

Description of Plate 110 ♀.—*a*, spider magnified; *b*, natural size; *c*, profile; *d*, cephalothorax underside; *e*, eyes; *f*, epigyne.

♀ Total length, 8, cephalothorax, 4, breadth, 2¾, in front, 1¼, abdomen, 5, breadth, 3 millim. Leg, i.—15, ii.—14¼, iii.—13½, iv.—12 millim. Patella + tibia, iv.—4 millim.

This spider is of a rich yellow colour with reddish and brown markings on back of abdomen.

Several specimens found at Singapore. ♂ Unknown.

Description of Plate 110.—Snares of Eurychoera quadri-maculata Thor.

No. 1. A. Reticulated snare below which was a platform of web seven inches long from E to F, B, mass of debris suspended by silken cords from A to B length of cords four inches. C, twisted leaf suspended in snare in which the spider was seated. When disturbed it retreated to D, leaf above the snare. G, loose leaf suspended from snare.

No. 2. Reticulated snare about twenty-four inches high in which was suspended a twisted leaf four inches long forming a nest for the spider. A platform of web below the snare twelve inches long.

No. 3. Reticulated snare about six inches high made among the leaves of a convolvulous plant. Spider was living on the lower leaf.

No. 4. Nest in the fork of two leaves, no reticulated snare. Spider carrying its cocoon between the falces and the fore legs.

The nest in twisted leaf in which the spider lives seems from my notes to be mostly tubular.

MATIDIA TRI-NOTATA THOR

1890. ***Matidia ? tri-notata*** Thor. Aracnidi di Pinang, p. 24.

Description of Plate 112 ♀.—*a.* and *aa*, spider magnified; *b*, natural size; *c*, profile; *d*, maxillae and labium; *e*, eyes; *f*, epigyne; *g*, abdomen underside.

♀ Total length, 4; cephalothorax, 2, length about 1⅓; abdomen rather more than 2; breadth about 1 millim. Leg i ? (about 4 ?), ii.—about 5, iii.—3½, iv.—nearly 5½ millim. Patella + tibia, iv.—almost 2 millim. (Thorell).

Cephalothorax of a brown colour, abdomen and legs yellow.

One specimen found in Singapore. Habits unknown.

FECENIA PROTENSA ? THOR.

1891. *Fecenia protensa*, Thor. Spindlar Nikobarerna, etc., p. 31.

Description of Plate 113 ♀.—*a*, spider magnified; *b*, natural size; *c*, profile; *d*, cephalothorax underside; *e*, eyes; *f*, epigyne; *g*, snare.

Total length, $9\frac{3}{4}$; cephalothorax, $3\frac{3}{4}$; breadth, $2\frac{1}{3}$; breadth in front, 2; abdomen, 6; breadth, $2\frac{4}{5}$ millim. Leg, i.—22; ii.—13; iii.—8; iv.—12 millim. Patella + tibia, iv.—4 millim. (Thorell.)

This species makes a circular snare, three inches in diameter, in the centre of which it lives in a nest made of debris or a twisted leaf. The ♂ is about as large as the ♀ and is similarly coloured; unfortunately I did not obtain mature specimens.

DICTIS DOMESTICA DOL.

1859. *Scytodes domestica*, Dol., Tweede Bijdr., etc., p. 48. Tab. 6. Fig. 1.

1891. *Dictis fumidia*, Thor., Spindlar fr. Nikobarerna, etc., p. 33.

1894. „ *domestica*, ibid., Arachnider fran Java, etc., p. 6.

Description of Plate 114 ♀.—*a*, spider magnified; *b*, natural size; *c*, profile; *d*, cephalothorax underside; *e*, eyes.

Total length, $4\frac{1}{2}$; cephalothorax, 2; breadth nearly $1\frac{1}{2}$; breadth of clypeus about $\frac{3}{4}$; abdomen, $2\frac{1}{2}$, breadth about 2 millim. Leg, i.—$8\frac{2}{3}$; ii.—$7\frac{1}{2}$; iii.—$5\frac{1}{6}$; iv.—$7\frac{1}{2}$ millim. Patella + tibia, iv.—nearly $2\frac{1}{2}$ millim. (Thorell.)

Dr. Chas. Aurivillius found one specimen in a termite's nest in the island of Edam, as noted in Thorell's Spindlar fran Java, etc., page 6.

THERIDIUM DELTAFORME N. Sp.

Description of Plate 115 ♀.—*a*, spider magnified; *b*, natural size; *c*, profile; *d*, cephalothorax underside; *e*, eyes; *f*, epigyne; *g*, snare ($\frac{1}{4}$ natural size).

Total length, 2; cephalothorax, 1; breadth, $\frac{3}{4}$: abdomen, $1\frac{1}{2}$; breadth, 2 millim. Leg, i.—$3\frac{1}{2}$; ii.—3; iii.—$2\frac{1}{2}$; iv.—$3\frac{1}{4}$ millim. Patella + tibia, iv.—$1\frac{1}{4}$ millim. Measurements only approximate.

Two specimens found at Singapore; one was climbing over a small web in a bush, and the other with its cocoon was in a beautiful reticulated web between two leaves.

THERIDIUM AUREUM N. Sp.

Description of Plate 116 ♀.—*a*, spider magnified; *b*, natural size; *c*, profile; *d*, cephalothorax underside; *e*, eyes; *f*, epigyne (wet).

Total length, 4½; cephalothorax, 1¾; breadth, 1¼; breadth in front, ¾; abdomen, 3½: breadth, 2¾ millim. Leg, i.—5½; ii.—4½; iii.—3; iv.—5 millim. Patella + tibia, iv.—1¾ millim.

One specimen found at Singapore in a reticulated snare, higher than broad, with twisted leaf in centre in which the spider sits.

THERIDIUM LABECULOSUM N. Sp.

Description of Plate 117 ♀.—*a*, spider magnified; *b*, natural size; *c*, profile; *d*, cephalothorax underside; *e*, eyes; *f*, epigyne (wet); *g*, nest with cocoon.

Total length, 3½; cephalothorax, 1¼; breadth, 1¼; breadth in front, ¾; abdomen, 2¼; breadth, 1¾ millim. Leg, i.—4; ii.—4; iii.—3; iv,—about 3¾ millim. Patella + tibia, iv.—1⅛ millim,

One specimen of this spider was found at Singapore and four specimens at Penang.

They make their reticulated, funnel-shaped nests on the face of a twisted leaf. The Singapore spider had a small leaf fastened in the reticulation and under it was the spider and its cocoon.

THERIDIUM BLANDUM CAMBR.

1882. *Coleosoma blandum Cambr.* Proc. Zool. Socy., London, p. 428, pl. XXIX, figg. 3a—af.

1896. „ „ Work. Spiders, vol. I., p. 60., pl. 60.

1896. *Theridion blandum* Sim. La Feuille d. Jeunes Naturalistes, Paris.

Description of Plate 118 ♀.—*a*, spider magnified; *b*, natural size: *c*, profile; *d*, cephalothorax underside; *e*, eyes; *f*, epigyne.

Total length, $2\frac{1}{4}$; cephalothorax, 1; breadth, $\frac{1}{2}$; do. in front, $\frac{1}{4}$; abdomen, $1\frac{1}{2}$; breadth, 1 millim. Leg, i.—about $2\frac{3}{4}$; ii.—about $2\frac{3}{4}$; iii.—about $1\frac{1}{2}$; iv.—about $2\frac{3}{4}$ millim. Patella + tibia, iv.—rather more than 1 millim.

The ♂ of this spider has already been figured in volume 1, plate 60 of this work.

This spider, which is common in the conservatories of the Jardin de Plantes, Paris, makes a small, simple, reticulated snare. Its cocoon, which is large, whitish, and fluffy, is carried by the ♀ suspended from her spinnerets. It has been found in Ceylon, Singapore, the Philipines, and Florida.

THERIDIUM AUREO-MACULATUM N. Sp'

Description of Plate 119 ♀.—*a*, spider magnified; *b*, natural size; *c*, profile; *d*, cephalothorax underside; *e*, eyes; *f*, epigyne; *g*, snare.

Total length, 3½; cephalothorax, 1¼: breadth, ¾; breadth in front, ¼; abdomen, 2; breadth, 1¾ millim. Leg, i.—3¾; ii.—3¼; iii.—3; iv.—3½ millim. Patella + tibia, iv.—1¼ millim.

A few specimens of this spider were found in the Botanic Gardens, Singapore. It suspends by a silken thread its bell-shaped nest, composed of little bits of stone and earth, in dark holes in walls.

The nest is about half-an-inch long and is held together by a slight web, and stayed from the mouth by numerous threads.

LINYPHIA RUGOSA N. Sp.

Description of Plate 120 ♀.—*a*, spider magnified; *b*, natural size; *c*, profile; *d*, cephalothorax underside; *e*, eyes; *f*, epigyne (wet); *g*, do. profile (wet).

Total length, 4½; cephalothorax, 2; breadth, ¾; abdomen, 2¾; breadth, 2½ millim. Leg, i.—wanting tarsus, 10½; ii.—10¾; iii.—7¼; iv.—9¼ millim. Patella + tibia, iv.—3 millim.

Only one specimen of this spider found; it was near a reticulated snare and was carrying its cocoon in its mouth.

ULOBORUS GENICULATUS Oliv.

1789.—Aranea geniculata, Oliv. Encycl. Méthod., etc., ii. p. 214.
1895.—Uloborus geniculatus, Thor. The Spiders of Burma, p. 127.

Description of Plate 121 ♀.—*a*, spider magnified; *b*, natural size; *c*, profile; *d*, cephalothorax underside; *e*, eyes; *f*, epigyne; *g*, do. profile; *h*, ♂ right palpus from below; *i*, right palpus in front; *k*, left palpus from outside.

This species, which is found in all tropical countries, makes an orbicular horizontal snare, in the centre of which it suspends itself back downwards, with the legs stretched out in front and behind.

ULOBORUS RAFFRAYI Sim.

1891.—Uloborus raffrayi Sim. Ann. Soc. Ent. France, p. 8, pl. iv.

Description of Plate 122 ♀ —*a*, spider magnified; *b*, natural size; *c*, profile; *d*, cephalothorax underside; *e*, eyes; *f*. ♂ right palpus from outside; *g*, left palpus in front; *h*, epigyne; *i*, epigyne profile.

Plate 123, snares of a colony of U. raffrayi, greatly reduced in size.

ULOBORUS SOLIDUS, n. sp.

Description of Plate 124 ♀.—*a*, spider magnified; *b*, natural size; *c*, profile; *d*, cephalothorax underside; *e*, eyes; *f*, epigyne (wet); *g*, do. profile (wet).

Total length, 5; cephalothorax, 1½; breadth, about 1; abdomen, 4; breadth, 2½ millim. Leg, 1—7; 2—3; 3—3; 4—5½ millim. Patella + tibia, iv.—2½ millim.

Only two specimens of this spider found at Singapore in a reticulated snare.

ARGYROEPEIRA BIVERTEX n. sp.

Description of Plate 125 ♀.—*a*, spider magnified; *b*, natural size; *c*, profile; *d*, cephalothorax underside; *e*, eyes; *f*, epigyne; *g*, snare.

This spider, of which I obtained three specimens, makes a horizontal snare in dark undergrowth, seven inches in diameter.

Rays, 30.
Inner spiral, 5 turns.
Free zone, ½".
Outer spiral, 70 turns.

There is a piece of debris hanging below the web, suspended by a line of white floss silk, which silken line goes out to the margin of the snare. When disturbed, the spider retreated to this piece of debris.

ARGYROEPEIRA NIGRO-TRIVITTATA, Dol.

1859.—Epeira nigro-trivittata, Dol. Tweede Bidr, etc., p. 39; tab. xi.; fig. v.
1881.—Meta nigro-trivittata, Thor. Studi, etc., iii., p. 126.
1889-90.—Argyroepeira do. id. do. iv., p. 196.

Description of Plate 126 ♀.—*a*, spider magnified; *b*, natural size; *c*, profile; *d*, cephalothorax underside; *e*. eyes; *f*, epigyne.

Total length, 11½; cephalothorax, 4; breadth, 3; do., in front a little more than 1½; abdomen, 9½; breadth. 5⅓ millim. Leg, i.—25¾; ii.—19½; iii.—10½; iv.—17 millim. Patella + tibia, iv.—5 millim. (Thor.).

NEPHILENGYS MALABARENSIS, Walck.

1841.—Epeira malabarensis, Walck, H.N. d. Ins. Apt., ii., p. 102.
1895.—Nephilengys do. Thor. The Spiders of Burma, p. 160.

Description of Plate 127 ♀.—*a*, spider magnified; *b*, natural size; *c*, profile; *d*, labium and falces; *e*, eyes; *f*, epigyne; *g*, ♂ palpus; *h*, two forms of ♂; *i*, size of smaller form; *k*, snare on face of wall; *l*, do. on tree (greatly reduced).

This large and handsome spider is found in all parts of the tropics. It is very variable in colour and size. Its snare is usually only the segment of a circle, and in most cases has a tunnel from the centre of the web into which it retreats. There are two distinct forms of the ♂. The smaller and commoner form is of a dark chestnut brown colour, while the other has markings similar to the ♀. This latter form I have only found in Colombo.

EUETRIA MOLUCCENSIS, Dol.

1857.—Epeira moluccensis, Dol. Bijdr., etc., xiii., p. 418; tab. 2a; fig. 6: tab. 22; fig. 6.
1895.—Euetria do. Thor. The Spiders of Burma, p. 169.

Description of Plate 128 ♀.—*a*, spider magnified; *b*, natural size; *c*, profile; *d*, cephalothorax underside; *e*, eyes; *f*, epigyne profile; *g*, do., in front.

HERENNIA ORNATISSIMA, Dol.

1859. Epeira ornatissima, Dol. Tweede Bidjr, etc., p. 32; tab. i.; fig. v.
1859. do. multipuncta id., ibid., p. 32; tab. xi.; fig. i.—ib.
1887. Herennia do. Thor. Studi., etc., p. 166.
1895. do. do. id. Spiders of Burma, p. 163.

Description of Plate 130 ♀.—*a*, spider magnified; *b*, natural size; *c*, profile; *d*, cephalothorax underside; *e*, eyes; *f*, epigyne; (dry) *g*, do. (wet).

I found a few specimens of this handsome spider running on the front of the Curator's House at the Botanic Gardens, Singapore, but did not notice their webs. Oates says it "makes a web about three feet long on a smooth tree trunk. Width ⅓ or ¼ of girth of tree. All the lines are vertical, forming a perfect rope ladder. The web follows the convexity of the trunk, and is everywhere about ½ inch from it. Verticals about 1 inch apart, horizontals about ¼ inch apart" (Thorell's Spiders of Burma page 163). Simon says in his Histoire Naturelle des Araignées page 758 that it makes an orbicular web.

ARGIOPE PAPUANA n. sp.

Description of Plate 129 ♀.—*a*, spider magnified; *b*, natural size; *c*, profile; *d*, cephalothorax underside; *e*, eyes; *f*, epigyne; *g*, do., profile.

Total length, 11; cephalothorax, 4½; breadth, 3½; breadth in front, 2; abdomen, 7; breadth, 4¼ millim. Leg, i.—23½; ii.—23; iii.—13¼; iv.—21½ millim. Patella + tibia, iv.—7 millim.

Bought from Janion, and noted as collected at Port Moresby, New Guinea, by H. O. Forbes.

POLTYS ILLEPIDUS C. L. Koch.

1843.—Poltys illepidus C. L. Koch Die Arachniden. Vol. 10, p. 97, fig 811.

Description of Plate 131 ♀.—*a*, spider magnified; *b*, natural size; *c*, profile; *d*, cephalothorax underside; *e*, eyes.

POLTYS APICULATUS Thor.

1892.—Poltys apiculatus Thor. Bul. Soc. Ento. Italiana xxiv. Tri. iii., p. 20.

Description of Plate 132 ♀.—*a*, spider magnified; *b*, natural size; *c*, profile; *d*, cephalothorax underside; *e*, eyes; *f*, epigyne profile (immature).

Total length, 11; cephalothorax, 5; breath, 3½; abdomen, 9; breadth, 6½ millim. Leg i.—17¾; ii.—17; iii.—11; iv.—13 millim. Patella + tibia iv.—5 millim.

This spider was found in a dauber wasp's nest at Singapore.

ARANEUS (EPEIRA) AUSTERUS n. sp.

Description of Plate 133 ♀.—*a*, spider magnified; *b*, natural size; *c*, profile; *d*, cephalothorax underside; *e*. eyes: *f*, epigyne; *g*, do. profile.

Total length, 4½; cephalothorax, 1½; breadth, 1¼; do. in front, 1; abdomen, 3¼; breadth, 4 millim. Leg, i.—4½; ii.—4¼; iii.—3¾; iv.—5¼ millim. Patella + tibia, iv.—1¾ millim.

Bought from Janion, and noted as collected at Ko Sula Bantam Java by H. O. Forbes.

ARANEUS RUSSATUS n. sp.

Description of Plate 134 ♀.—*a*, spider magnified; *b*, natural size; *c*, profile; *d*, cephalothorax underside; *e*, eyes: *f*, epigyne.

Total length, 7; cephalothorax, 2½; breadth, 2¼; do. in front, 1; abdomen, 5; breadth, 4 millim. Leg i.—8; ii.—?; iii.—5½: iv.—9 millim. Patella + tibia iv.—3 millim.

ARANEUS PALLIDUS n. sp.

Description of Plate 135 ♀.—*a*, spider magnified; *b*, natural size; *c*, profile; *d*, cephalothorax underside; *e*, eyes; *f*, epigyne.

Total length, 4½; cephalothorax, 2; breadth, 1¼; breadth in front, ¾; abdomen, 3; breadth, 2 millim. Leg i.—9½: ii.—5; iii.—2½; iv.—4 millim. Patella + tibia iv.—1½ millim.

ARANEUS RUSTICUS n. sp.

Description of Plate 136 ♀.—*a*, spider magnified; *b*, natural size; *c*, profile; *d*, cephalothorax underside; *e*, eyes; *f*, epigyne; *g*, do., profile.

Total length, 6; cephalothorax, 2¼; breadth, 2; breadth in front, 1¼; abdomen, 4; breadth, 3 millim. Leg i.—8¼; ii.—7½, iii.—3¾; iv.—6½ millim. Patella + tibia iv—2¼ millim.

ARANEUS NOVELLUS n. sp.

Description of Plate 137 ♀.—*a*, spider magnified ; *b*, natural size ; *c*, profile ; *d*, cephalothorax underside ; *e*, eyes ; *f*, epigyne and spinnarets.

Total length, 3¼ ; cephalothorax, 1½ ; breadth about 1¾ ; in front, 1 ; abdomen, 2½ ; breadth, 2½ millim. Leg i.—4¾ ; ii.—4½ ; iii.—2½ ; iv.—3½ millim. Patella + tibia iv.—1½ millim.

Between the epigyne and the spinnarets are two large yellow spots.

Two specimens of this little spider were found by me at Singapore, one of them in an orbicular snare 9 inches in diameter, consisting of about 30 rays and 30 spirals. Spider was sitting in centre of web.

ARANEUS THEISII WALCK.

1841.—Epeira theisii Walck H. N., d. Ins., Apt. ii., p. 53 Atlas pl. 18, fig 4.

Description of Plate 138 ♀.—*a*, spider magnified: *b*, natural size; *c*, profile; *d*, cephalothorax underside; *e*, eyes.

This spider makes an orbicular vertical snare in long grass, and sits in a nest in the grass outside the snare, with forefeet on one of the rays (no trap line).

Snare No. 1, 10 inches in diameter.

Rays 24, meeting in ¼-inch circle.
Inner spiral, 4 rounds.
Free zone, ½-inch
Outer spiral, 25 rounds at upper side.
" 30 do. at lower side.

Snare No. 2, 6 inches in diameter.

Rays, 28.
Inner spiral, 4 rounds.
Outer spiral, 27 do. at upper side.
" 40 do. at lower side.

ARANEUS MELANOCRANIUS Thor.

1887.—Epeira melanocrania Thor., Ann. Mus., Geneva, xxv. p. 209.

Description of Plate 139 ♀.—*a*, spider magnified; *b*, natural size; *c*, profile; *d*, cephalothorax underside; *e*, eyes.

ARANEUS TARDIPES Thor.

1898.—Epeira tardipes Thor. Spiders of Burma, p. 193.

Description of Plate 140 ♀.—*a*, spider magnified; *b*, natural size; *c*, profile; *d*, cephalothorax underside; *e*, eyes; *f*, epigyne; *g*, snare.

This spider makes a circular vertical snare, five inches in diameter, horizontally across which is a roll of brown silk, with a break in the centre in which the spider sits.

Rays, 72.
Inner spiral, 6 rounds.
Free zone, ¼-inch.
Outer spiral, 53 rounds.

ARANEUS BANTAMENSIS n. sp.

Description of Plate 141 ♀.—*a*, spider magnified; *b*, natural size; *c*, profile; *d*, cephalothorax underside; *e*, eyes; *f*, epigyne; *g*, do. profile.

Total length, 5½; cephalothorax, 2; breadth, 2; abdomen, 5; breadth, 3 millim. Leg i.—?; ii.—7; iii.—5; iv.—6¼ millim. Patella + tibia iv.—2¼ millim.

ARANEUS OPHIR n. sp.

Description of Plate 142 ♀.—*a*, spider magnified; *b*, natural size; *c*, cephalothorax underside; *e*, eyes; *f*, epigyne.

Total length, 5½; cephalothorax, 2¾; breadth, 1½; abdomen, 2¾; breadth, 2½ millim. Leg i.—9; ii.—7; iii.—5; iv.—6 millim. Patella + tibia iv.—2 millim.

Two specimens found at Singapore, one in a vertical snare 7 inches in diameter,

Rays, 32.
Inner spiral, 6 rounds.
Free zone, ¾-inch.
Outer spiral, 37 rounds.

The other in an almost horizontal snare 10 inches in diameter, in the centre of which on the lower side the spider was sitting.

Rays, 38.
Inner spiral, 6 rounds.
Free zone, 1-inch.
Outer spiral, 57 rounds.

ARANEUS SCURRILIS n. sp.

Description of Plate 143 ♂.—*a*, spider magnified; *b*, natural size; *c*, profile; *d*, cephalothorax underside; *e*, eyes; *f*, left palpus from beneath; *g*, right palpus from above.

Total length, 3½; cephalothorax, 2; breadth, 1¾; abdomen, 2; breadth, 1½ millim. Leg i.—5; ii.—4¾; iii.—2¼; iv.—3 millim. Patella + tibia iv.—1¼ millim.

ARANEUS MALANGENSIS n. sp.

Description of Plate 144 ♂.—*a*, spider magnified; *b*, natural size; *c*, profile; *d*, cephalothorax underside; *e*, eyes; *f*, palpus.

Total length, 3½; cephalothorax, 2; breadth, 1¼; abdomen, 2; breadth, 1¾ millim. Leg i.—8¾; ii.—7¼; iii.—3¾; iv.—5 millim. Patella + tibia iv.—2 millim.

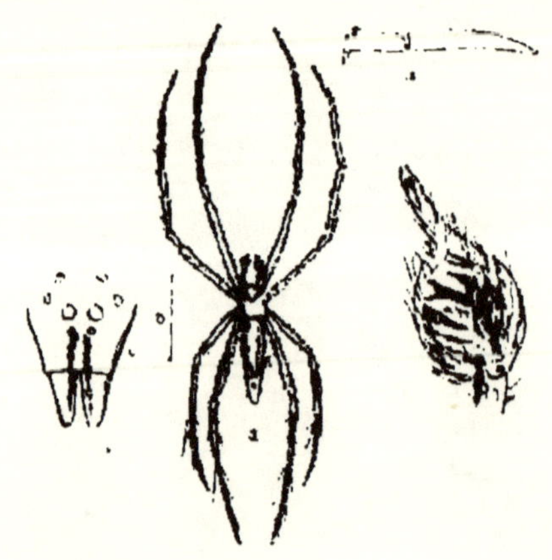

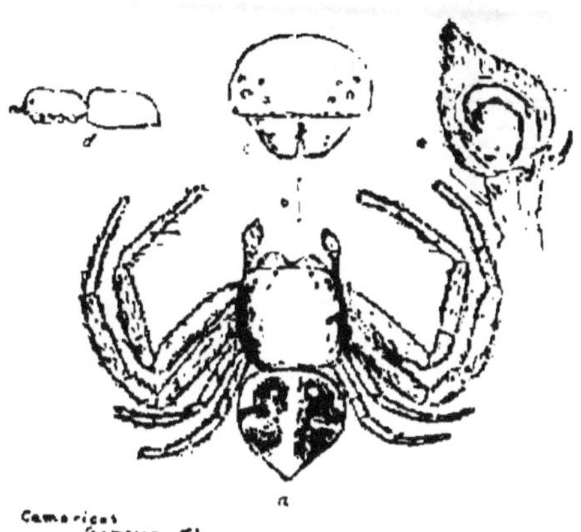

Camaricus
formosus Thor.
Penang

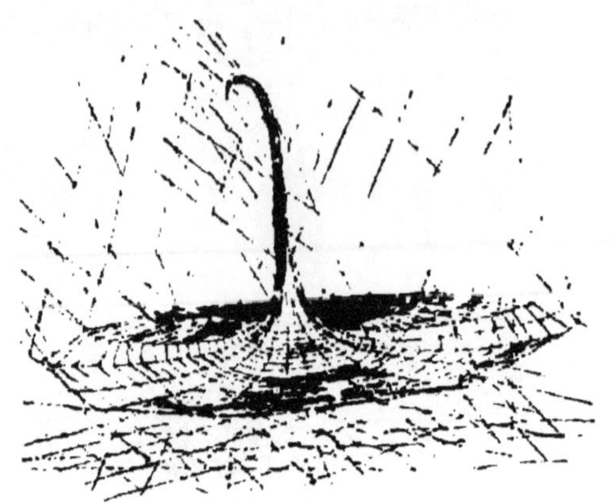

Simon

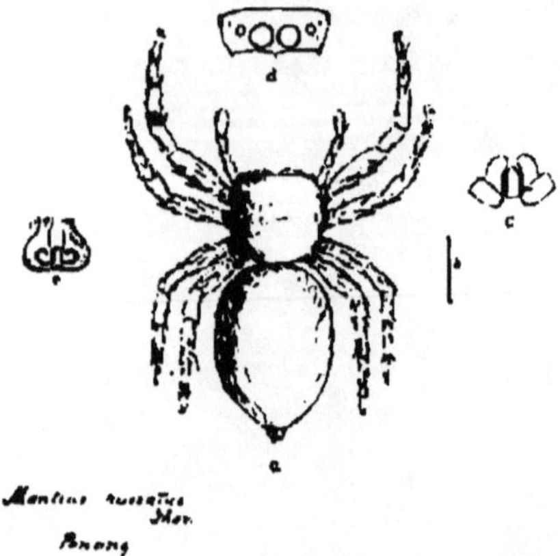
d
c
b
e
a
Mantius russatus
Thor.
Penang

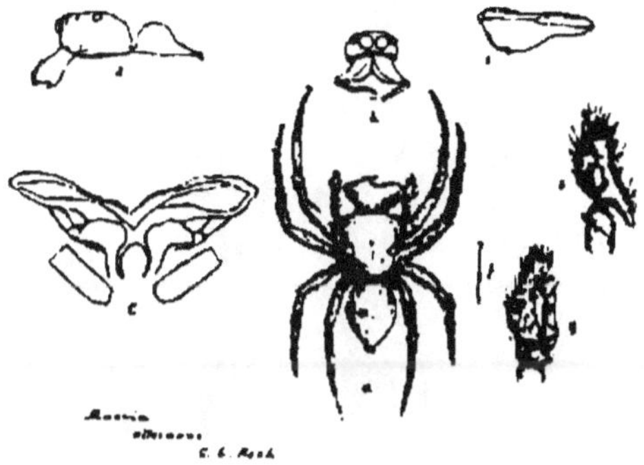

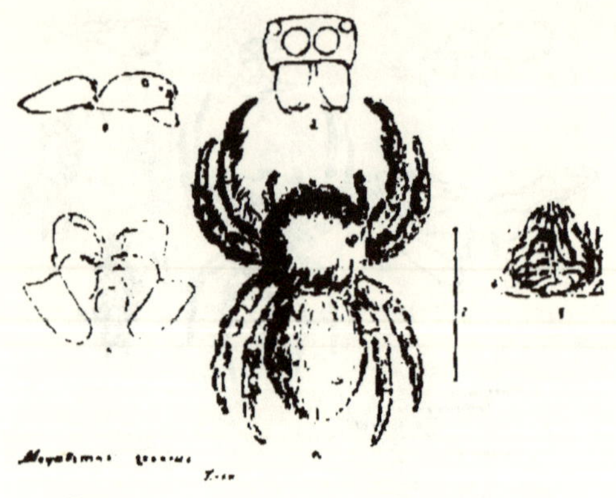
Penang

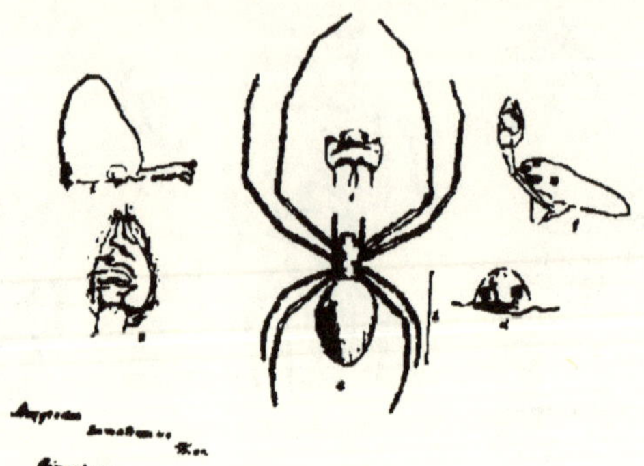

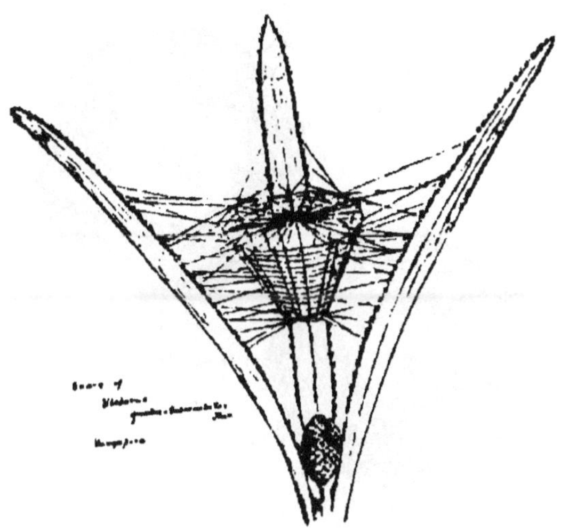

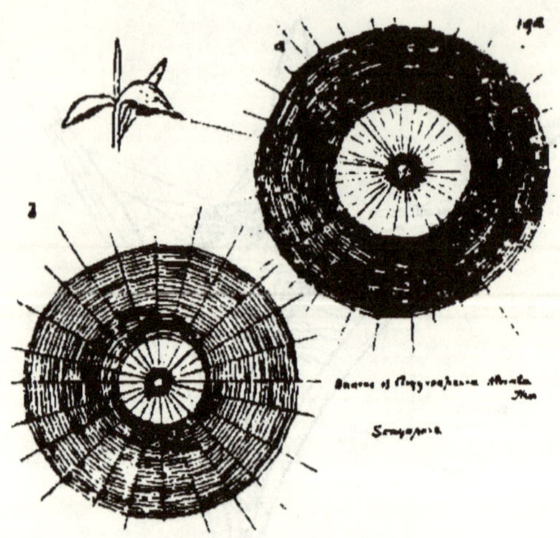

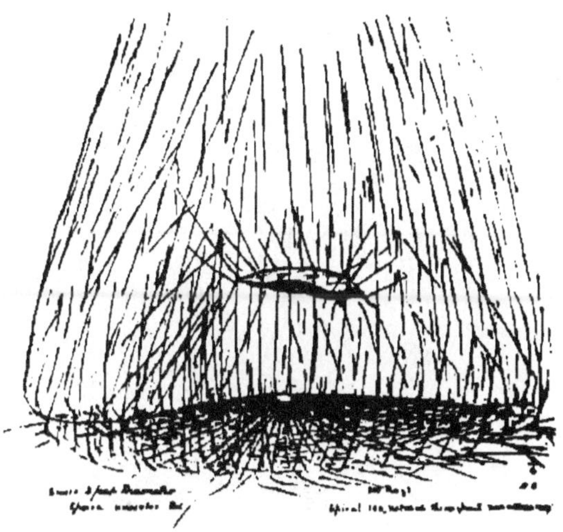

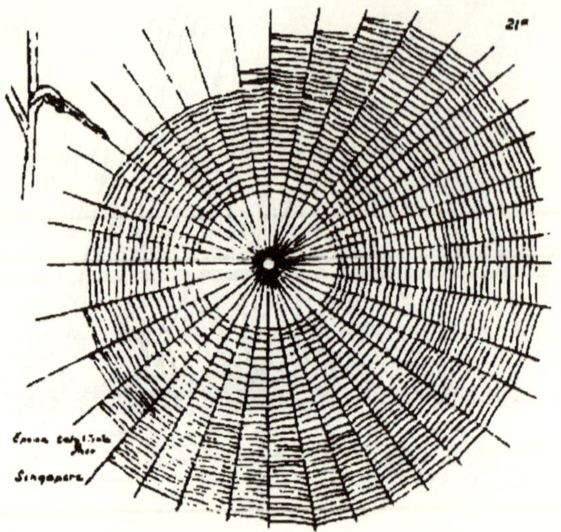
21a
Singapore

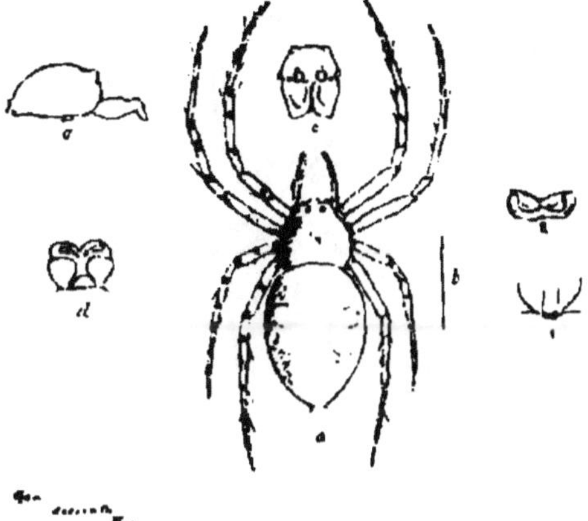
Singapore

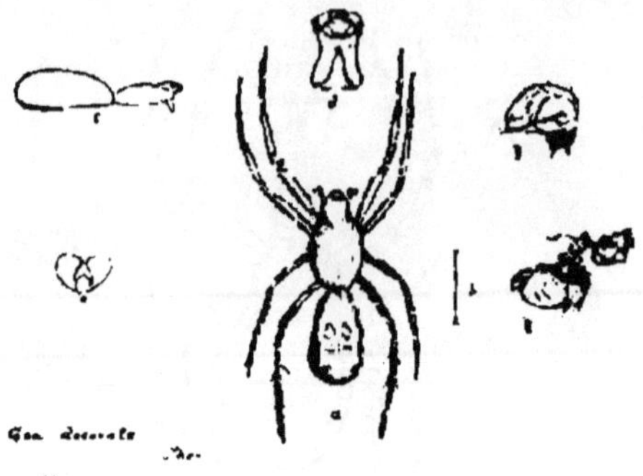

Gea decorata

Singapore

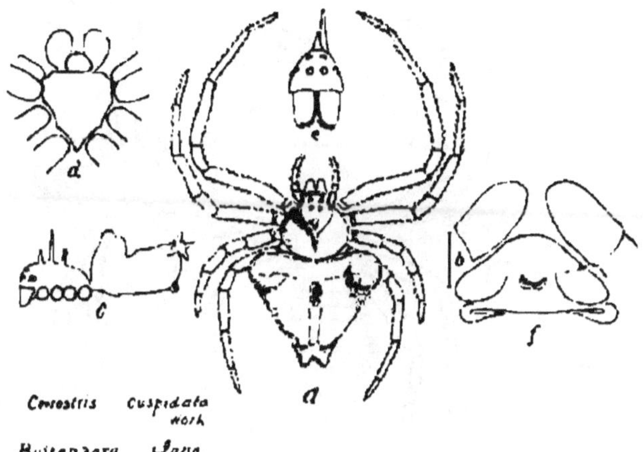

Cerostris cuspidata Work

Buitenzorg Java

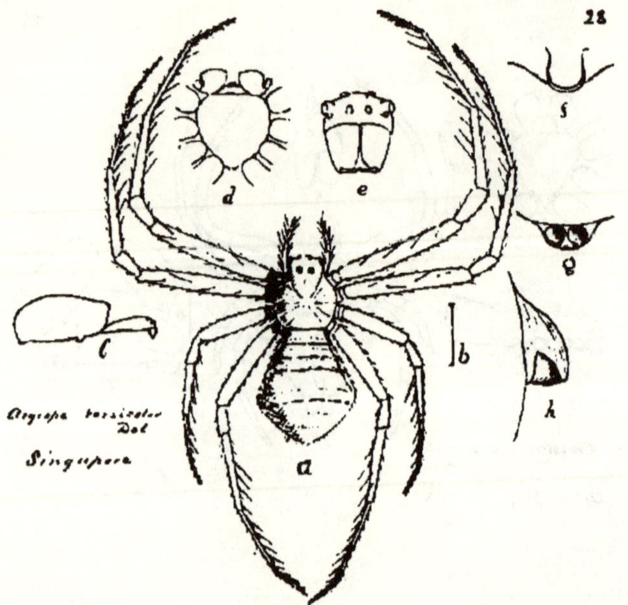
28
d
e
f
g
c
b
a
h
Argiope versicolor
Dol
Singapore

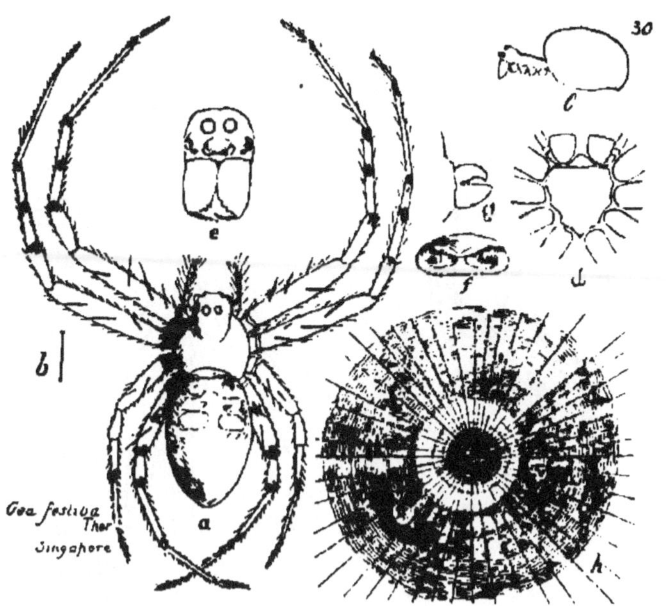
30
c
e
e
f
d
b
Gea festiva
Thor
Singapore
a
h

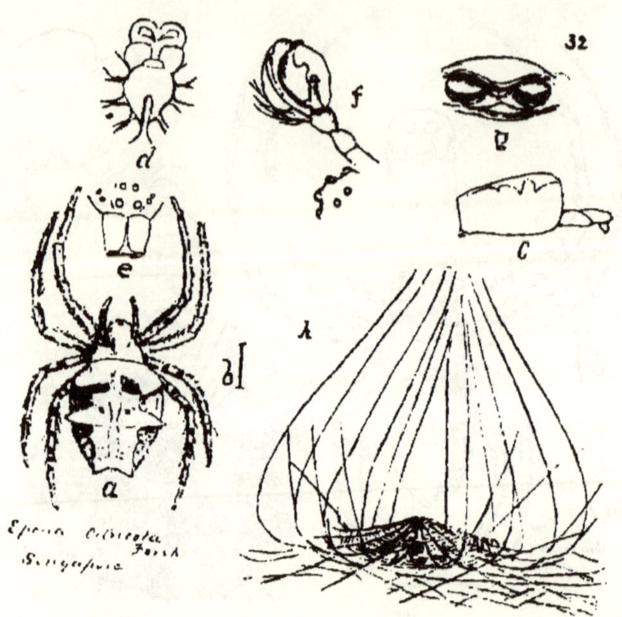
32
d
f
e
c
h
b
a

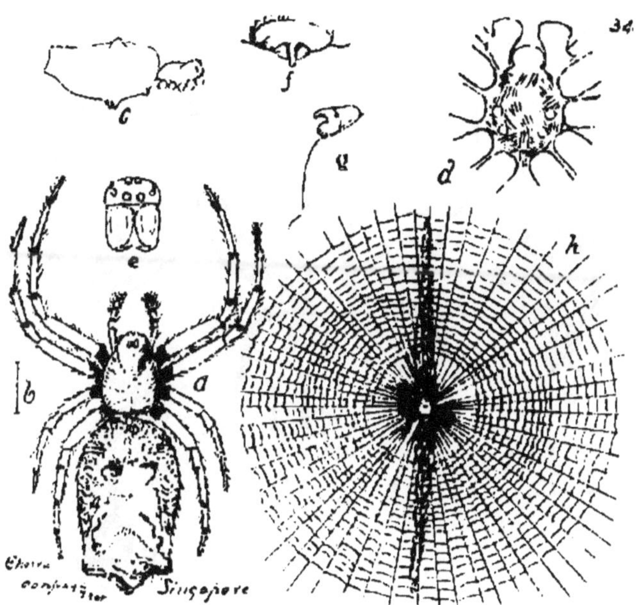
34.
c
f
g
d
e
h
a
b
Singapore

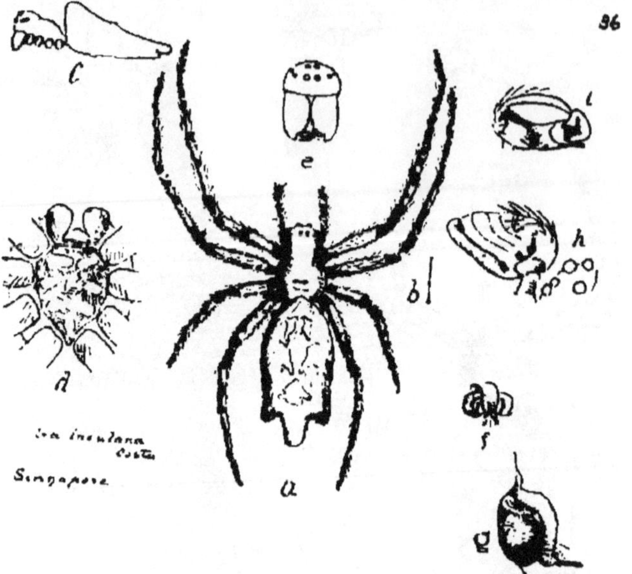
c
e
i
d
h
b
f
a
g
insulana
Costa
Singapore

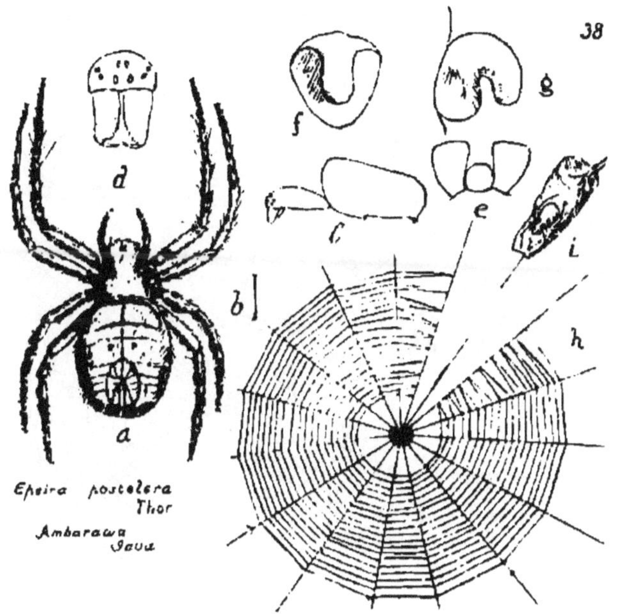
38
d
f
g
c
e
i
b
h
a
Epeira postelera
Thor
Ambarawa
Java

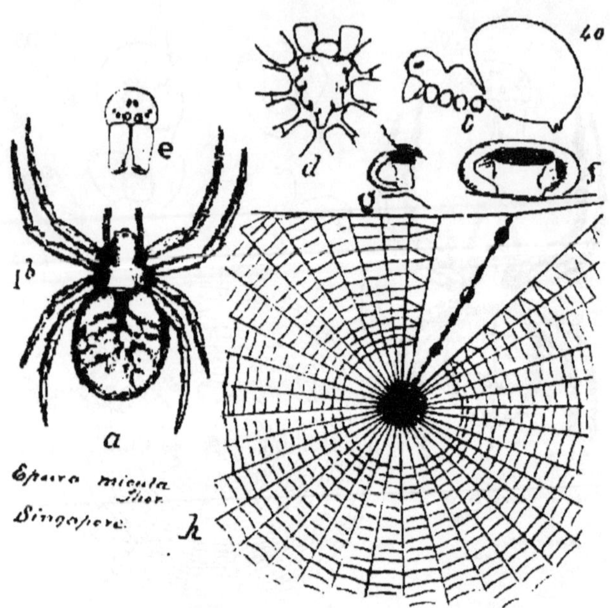

40
e
d
c
f
g
1b
a
Epeira micula
Thor.
Singapore.
h

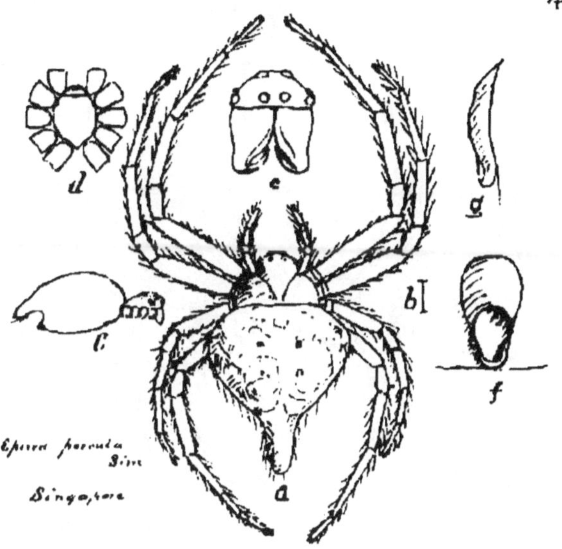
d
e
g
b
C
f
a
Epeira parvula
Sim
Singapore

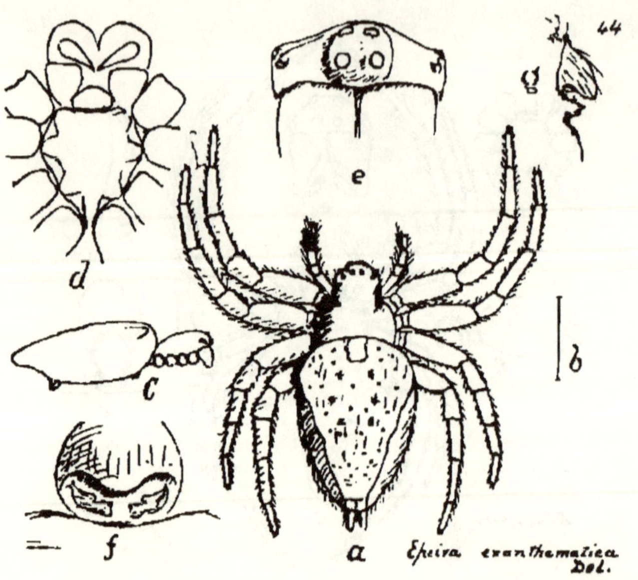
44
d
e
g
b
c
f
a
Epeira exanthematica
Dol.

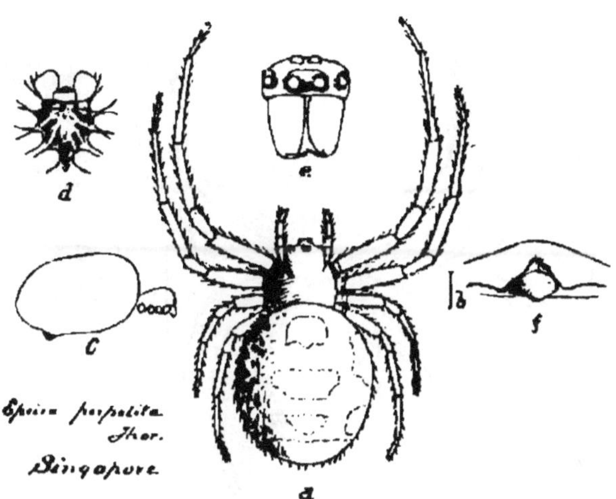
d
e
c
b
f
a
Epeira perpolita
Thor.
Singapore

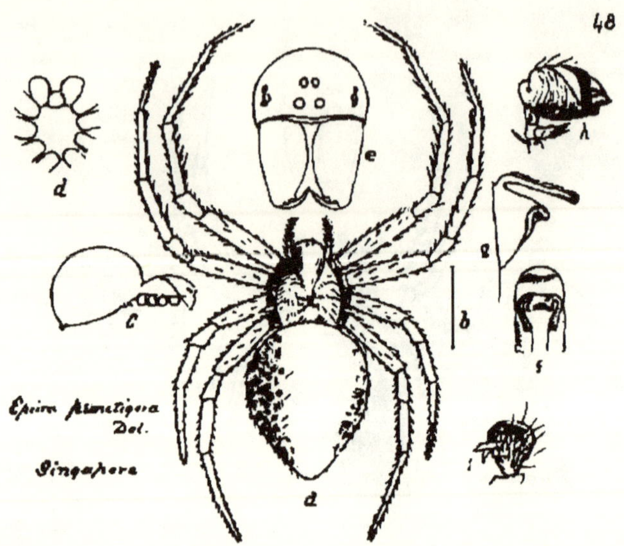
48
d
e
h
g
C
b
f
d
i
Epeira punctigera
Dol.
Singapore

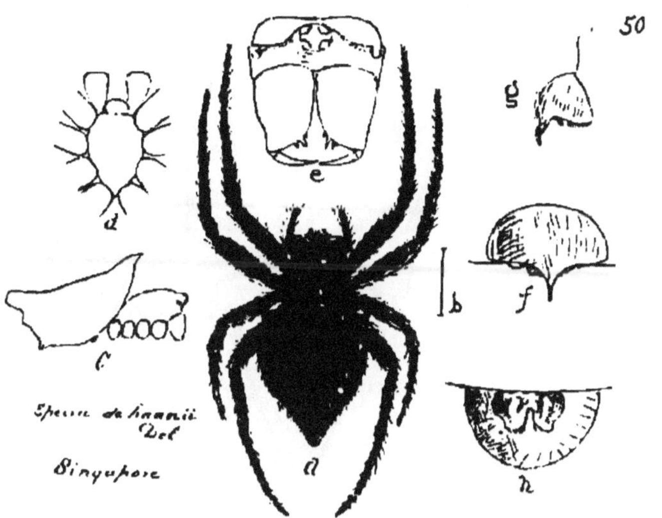

50
a
b
c
d
e
f
g
h
Singapore

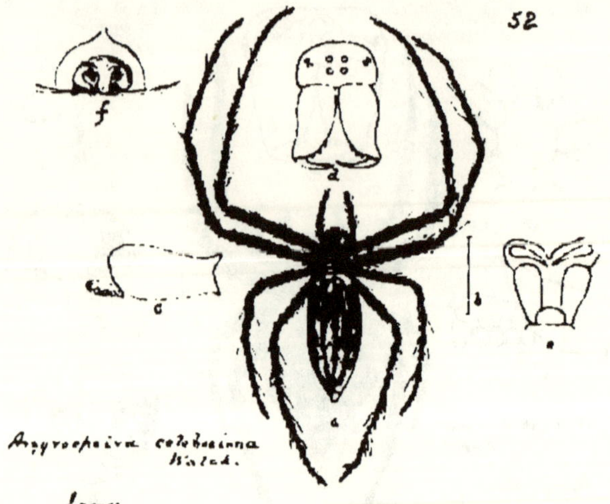
52
f
d
c
b
e
a
Argyroepeira celebesiana
Walck.
Java

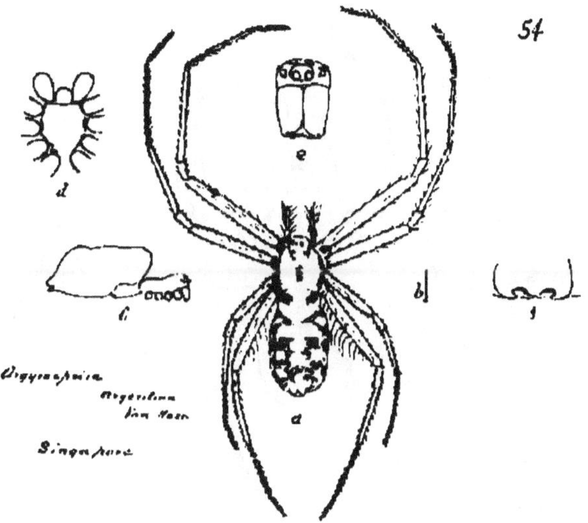
54
d
e
c
b
f
a
Singapore

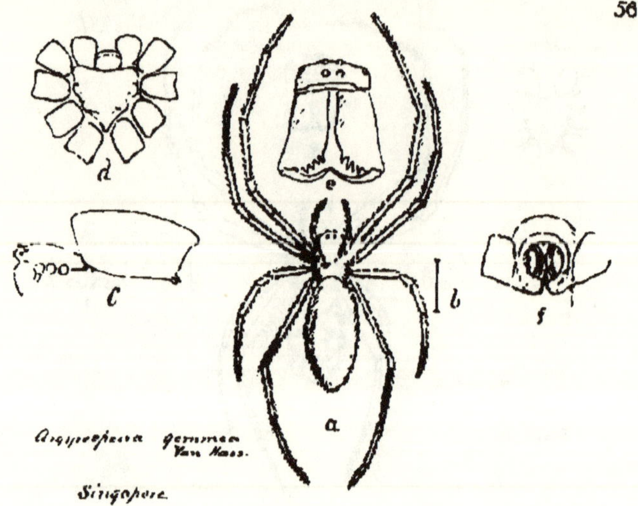
d
e
c
b
f
a
Argyroepeira gemmea
Van Hass.
Singapore

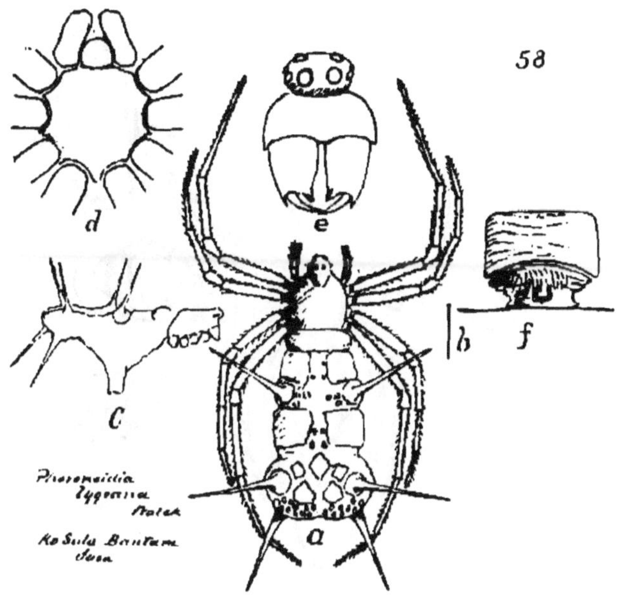
58
d
e
f
b
C
a

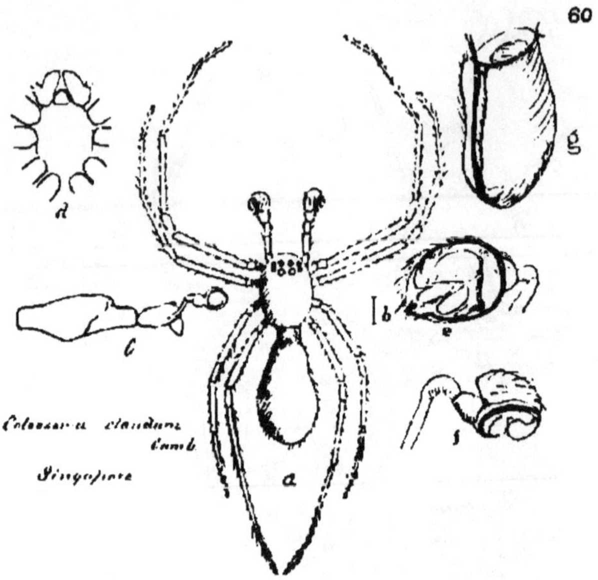
60
g
d
b
e
c
f
Camb.
Singapore
a

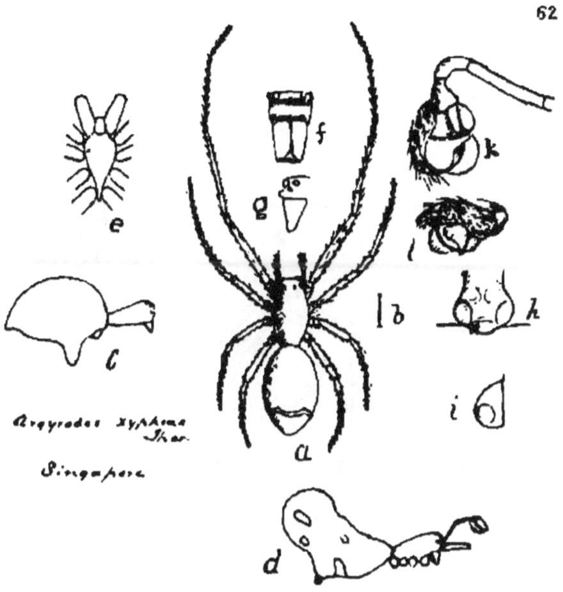
f
k
e
g
l
b
h
c
i
Argyrodes xyphema
Thor.
a
Singapore
d

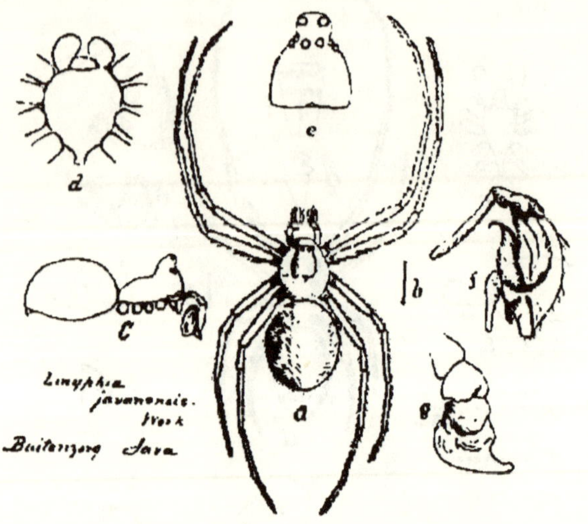
d
e
C
b
f
a
g
Linyphia
javanensis.
Strand
Buitenzorg Java

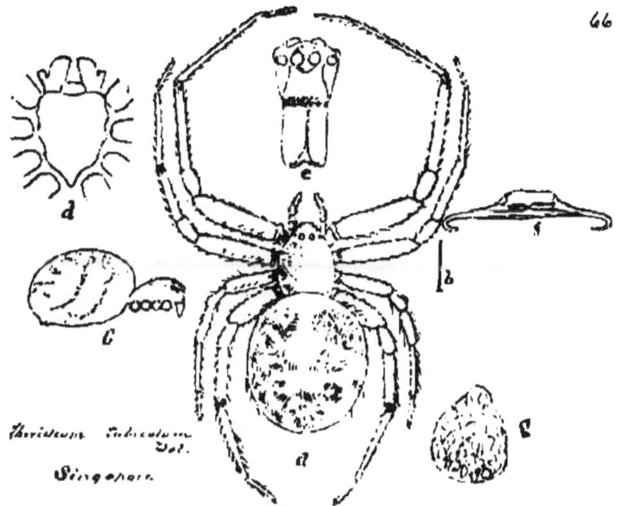
Singapore

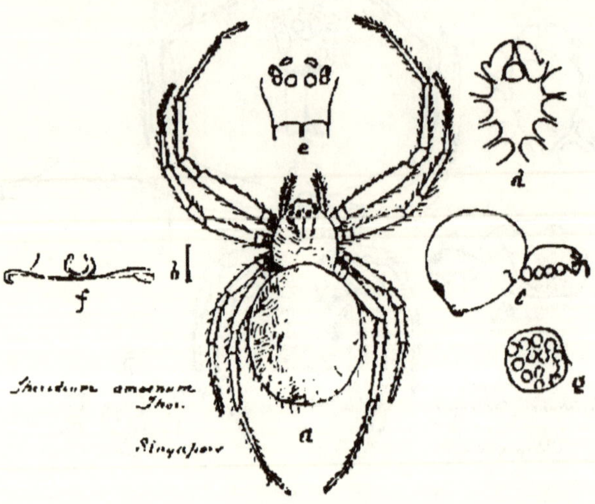
e
d
h
f
c
g
Theridium amoenum Thor.
a
Singapore

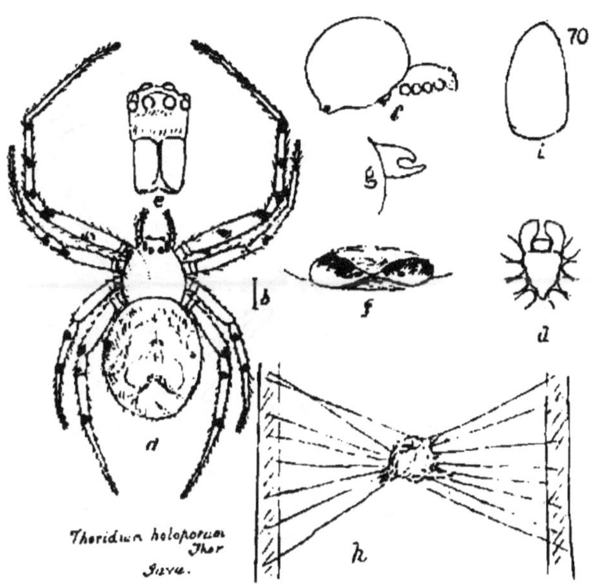
70
e
c
i
g
b
f
d
a
h
Theridium holoporum
Thor

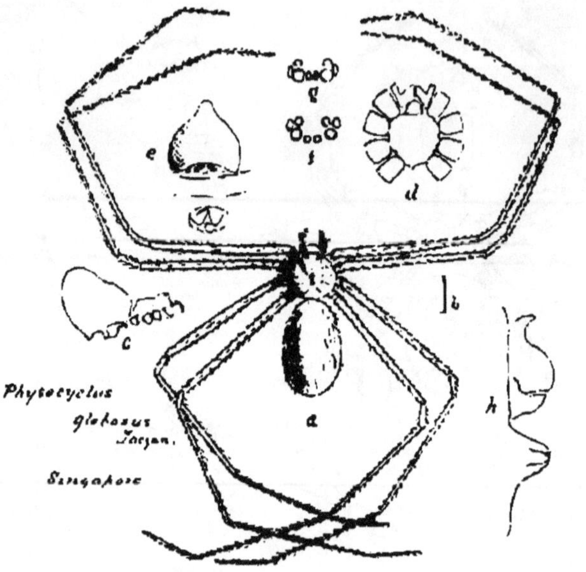

g
f
e
d
b
c
a
h
Phytocyclus
globosus
Jarjan.
Singapore

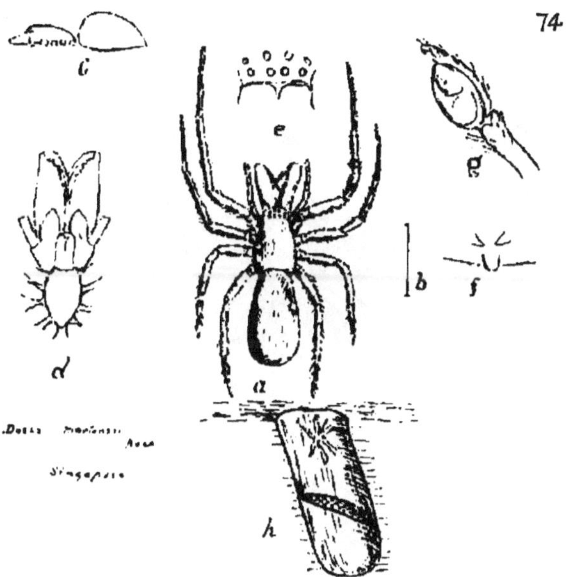
c
e
g
d
a
b
f
h
Singapore

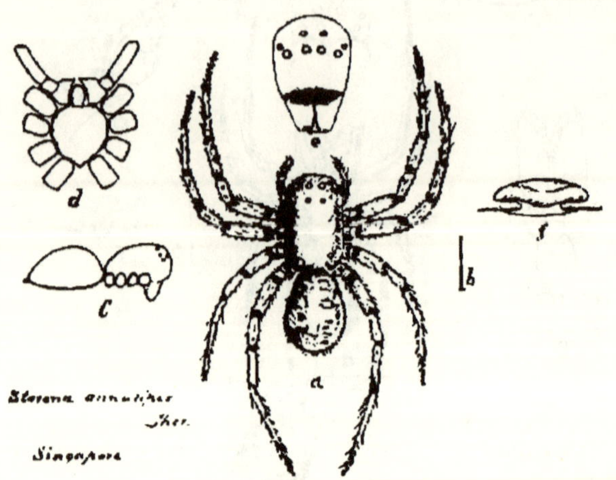

Storena annulipes Thor.

Singapore

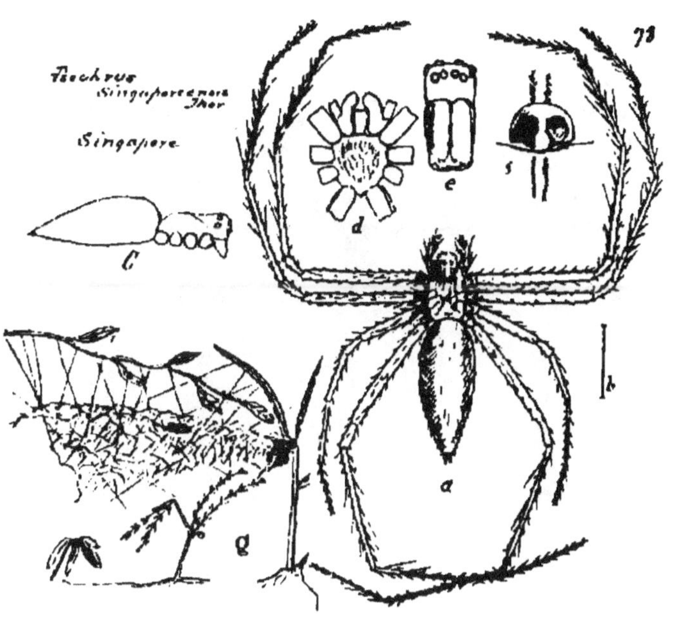
78
Psechrus
Singaporensis
Thor
Singapore
c
d
e
a
b
g

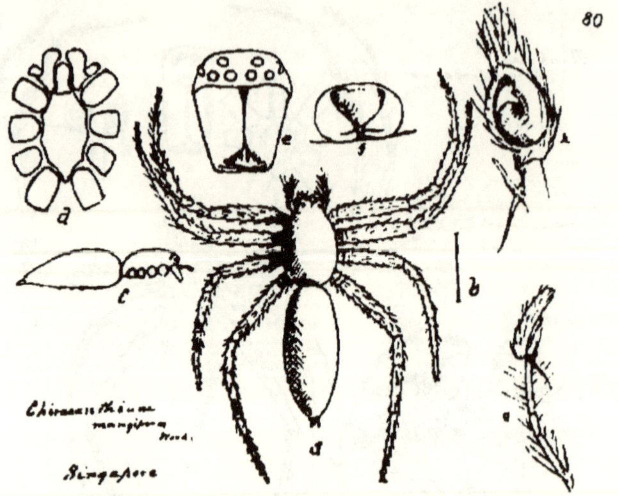
Chiracanthium
mangifera
Singapore

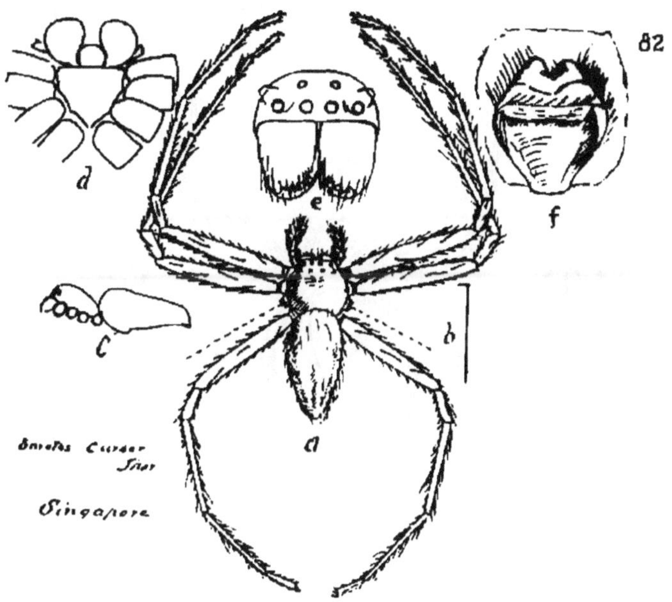
82
d
e
f
c
b
a
Singapore

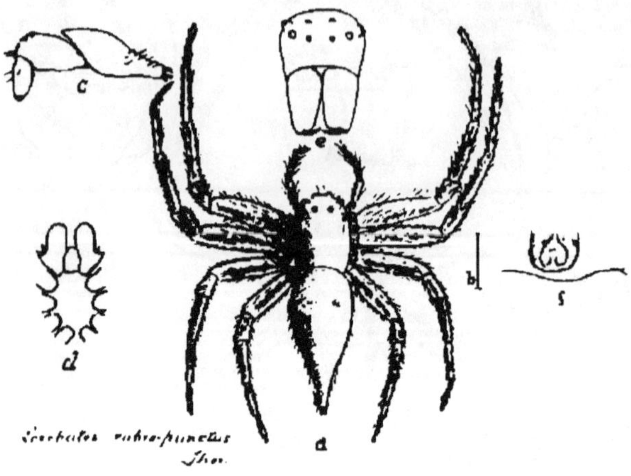

Cerbalus rubro-punctus Thor.

Singapore

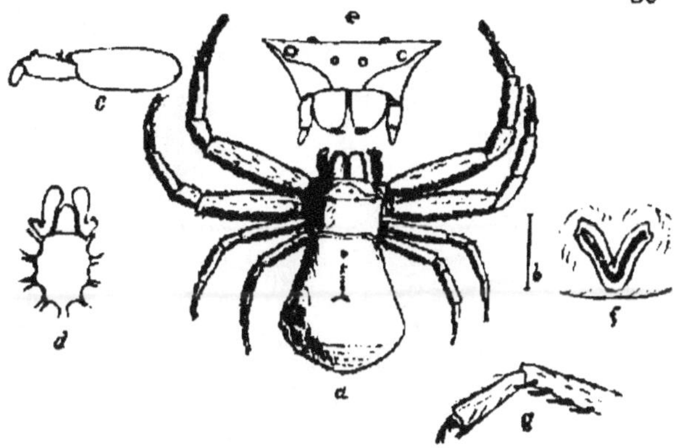

Daradius callidus Thor
Singapore

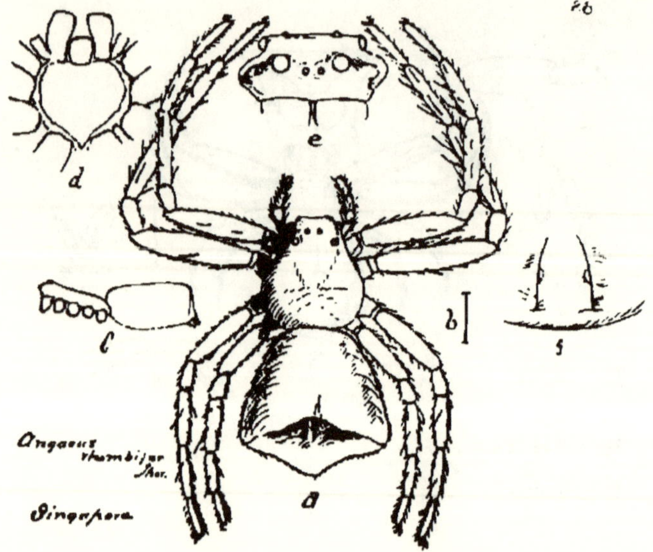
28
d
e
b
c
f
a
Angaeus rhombifer Thor.
Singapore

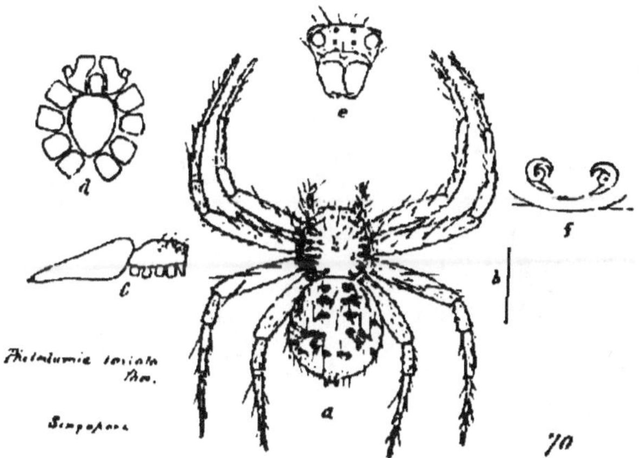
e
d
c
b
f
a
70

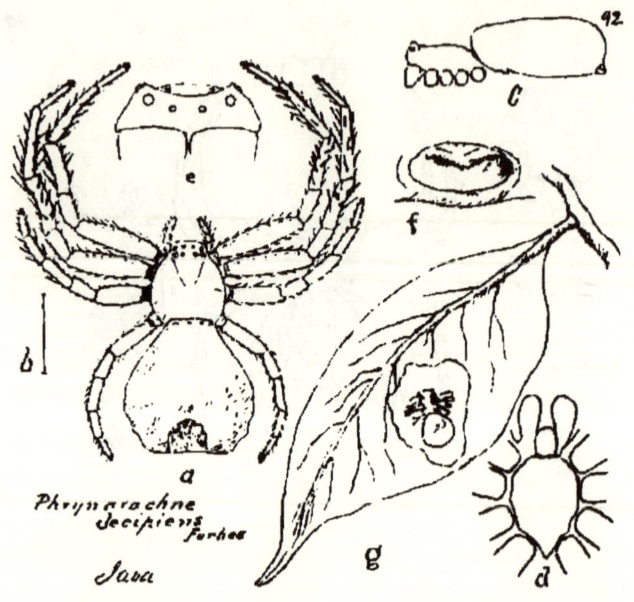
92
c
e
f
b
a
g
d
Phrynarachne
decipiens
Forbes
Java

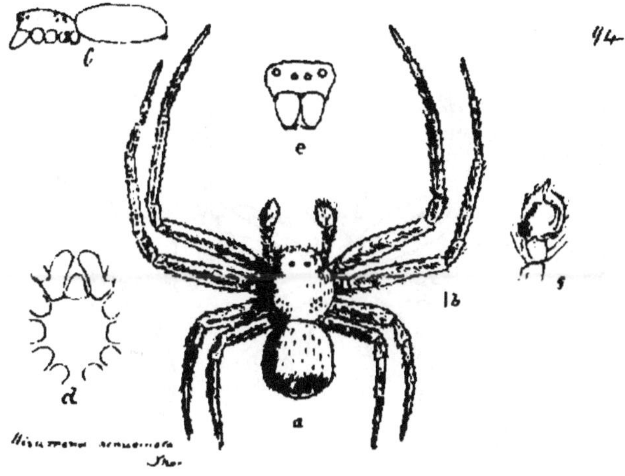

Singapore

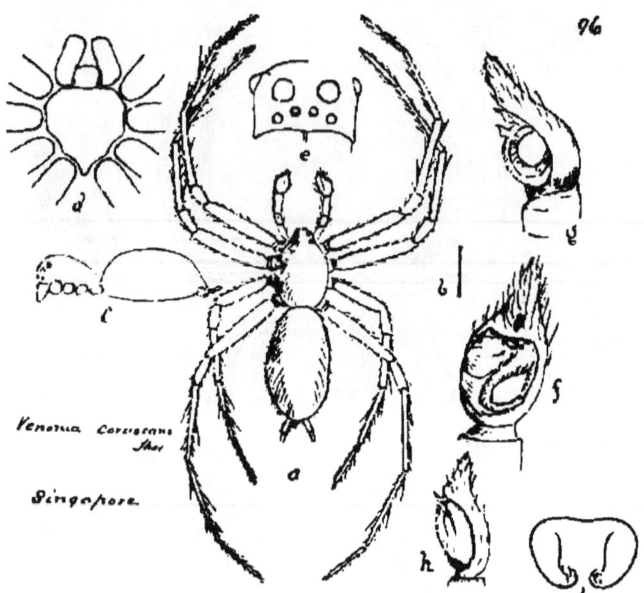
96
a
b
c
d
e
f
g
h
i
Venonia
Thor
Singapore

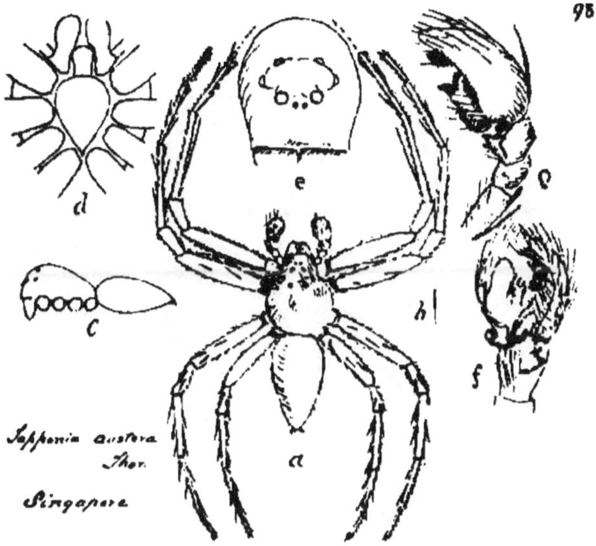
d
e
g
c
h
f
a
Sapponia austera
Thor.
Singapore

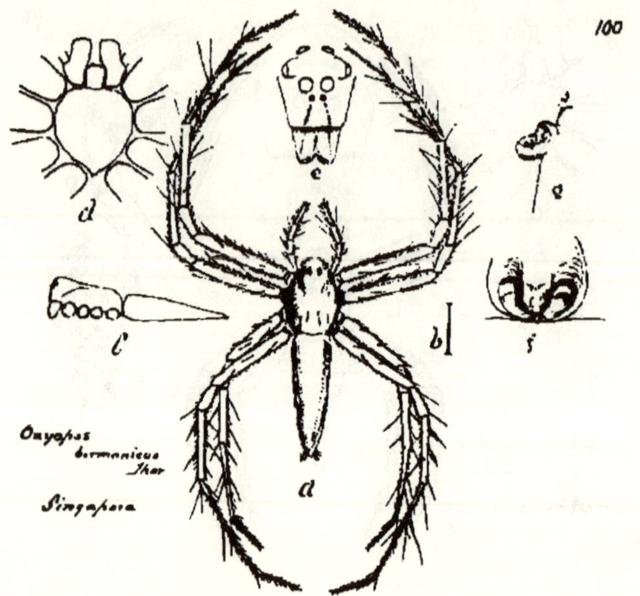
100
a
b
c
d
e
f
Oxyopes
birmanicus
Thor
Singapora

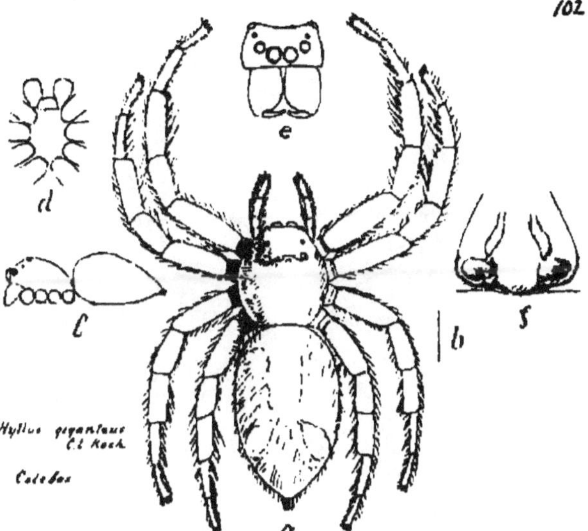
e
d
c
b
f
Hyllus giganteus
C. L. Koch
Celebes
a

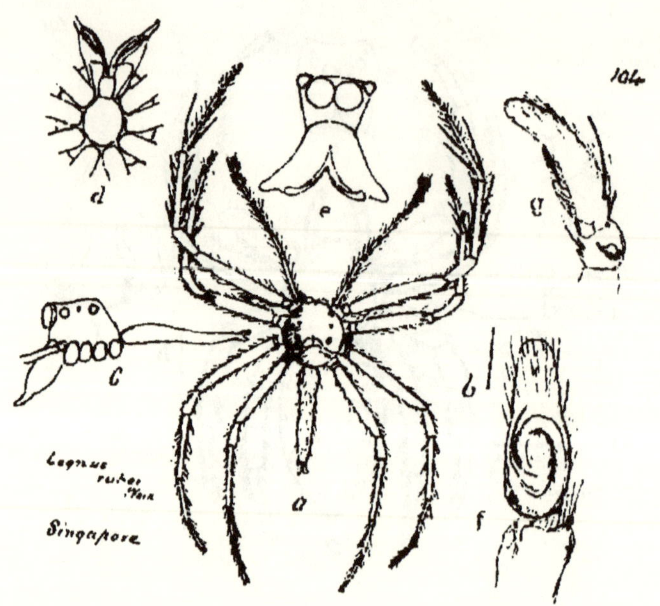

104
d
e
g
c
b
a
f
Singapore

www.ingramcontent.com/pod-product-compliance
Lightning Source LLC
Chambersburg PA
CBHW021504210326
41599CB00012B/1128

* 9 7 8 3 3 3 7 2 9 5 0 8 0 *